Architecture
木 建 筑

ISSUE 02　第 2 辑
AUTUMN 秋 2020
刘 杰 主编

中国建筑工业出版社

图书在版编目（CIP）数据

木建筑. 第2辑：汉英对照 / 刘杰主编. —北京：中国建筑工业出版社，2020.3
ISBN 978-7-112-24892-6

Ⅰ.①木… Ⅱ.①刘… Ⅲ.①木结构-建筑设计-文集-汉、英 Ⅳ.①TU366.2-53

中国版本图书馆CIP数据核字（2020）第032984号

责任编辑：李 鸽 陈小娟 陈海娇
责任校对：芦欣甜

木建筑

ISSUE 02　　第2辑
AUTUMN 秋 2020

刘 杰 主编

*

中国建筑工业出版社出版、发行（北京海淀三里河路9号）
各地新华书店、建筑书店经销
北京雅盈中佳图文设计公司制版
北京雅昌艺术印刷有限公司印刷

*

开本：965×1270毫米　1/16　印张：19　字数：588千字
2020年10月第一版　2020年10月第一次印刷
定价：480.00元
ISBN 978-7-112-24892-6
（35640）

版权所有　翻印必究
如有印装质量问题，可寄本社退换
（邮政编码 100037）

Mù
Architecture
木建筑

▲ ISSUE 02　第2辑
▲ AUTUMN 秋 2020

● 大跨度木结构建筑专辑 ●
Special Issue for Large-span Timber Architecture

主办单位：　上海交通大学设计学院木建筑研究与设计中心
支持单位：　上海交通大学　欧洲木业协会　加拿大木业协会
学术顾问：　（按姓氏笔画排序）
　　　　　　王建国　沈世钊　程泰宁

主　　编：　刘杰
编委会委员：（按姓氏笔画排序）
　　　　　　王兴田　阮昕（澳大利亚）李保峰
　　　　　　约根·赫曼松（瑞典）杨学兵　扬·索德林（瑞典）
　　　　　　张绍明　赵川　赵辰　黄华力（中国香港）
　　　　　　斯特凡·温特（德国）赫尔曼·考夫曼（奥地利）
编辑部主任：沈姗姗　李鸽
编辑部成员：（按姓氏笔画排序）
　　　　　　东鸿（美国）　高瑜　陶亮　蒋音成　韩佳纹

Host: The Center for Timber Architecture Research & Design
Supporter: Shanghai Jiao Tong University
European Wood, Canada Wood
Consultant: WANG Jianguo, SHEN Shizhao, CHENG Taining

Chief Editor: LIU Jie
Editorial Committee: WANG Xingtian, RUAN Xing (Australia)
LI Baofeng, Jorgen Hermansson (Sweden)
YANG Xuebing, Jan Soderlind (Sweden)
ZHANG Shaoming, ZHAO Chuan
ZHAO Chen, Eric Wong
Stefan Winter (Germany)
Hermann Kaufmann (Austria)
Editorial Director: SHEN Shanshan, LI Ge
Executive Director: Abraham Zamcheck (USA)
GAO Yu, TAO Liang, JIANG Yincheng, HAN Jiawen

受到国家重点研发计划"绿色生态木竹结构体系研究及示范应用"项目资
课题名称：大跨木结构体系研究及工程示范　课题编号：2017YFC0703506
book is supported by an initiative of the National Key R&D Program of China
esearch and Demonstration Application of Green Ecological Wood Bamboo
cture Systems which is named "Large-span timber structure system research
and construction demonstration (No. 2017YFC0703506)".

主编的话
Editor's notes

18世纪中叶欧洲工业革命以来，随着钢铁与水泥产量的逐步增长，大跨度公共建筑和交通建筑也逐渐被钢铁和钢筋水泥化了。一段时期以来，在全球范围内钢铁的质与量几乎成了衡量一个国家工业化水平的标准。在建筑业内，钢铁裹挟着水泥更是成为主体建筑结构的不二选择。即使在今天，我们的建筑业依然没办法完全摆脱钢铁加水泥的材料组合。尽管钢铁和水泥的生产与人类生存环境的可持续发展存在着不可调和的矛盾，但从目前的科技水平和制造业基础来看，要彻底淘汰二者还是一个不切实际的幻想。因此，为了我们居住的这个星球可以更加健康长寿，建筑工业就不得不面临这样一个挑战：必须要求更多的建造选择对环境有益的，且可持续发展的建筑结构材料。现代木竹结构建筑体系恰恰是人类在面临日益增长的环境危机背景下积极探索出来的一个解决方案。

长期以来，建筑结构发展的目标主要是朝着两个维度展开，一个是高度——设计师与建造者总是希望建筑可以变得更高，让建筑更加挺拔醒目的同时还可以节省珍贵的土地资源；一个是跨度——设计师与建造者总是希望建筑的跨度变得更大，可以创造更加宽敞的室内空间以及营造雄伟恢宏的空间效果。让建筑变得高度更高和跨度更大的主要结构性材料功臣——毫无悬念的都是钢铁和钢筋水泥。直至今日，在科学技术与环保理念的共同推动下，现代木竹结构材料体系也当仁不让地加入了争夺这两项桂冠的行列。令人可喜的是，在最近的十年里，新的可持续的结构性材料并没有令人失望，在高度上已经不断刷新着现代木竹结构建筑的纪录，在跨度上也在继承传统木结构建筑的基础上，有了新的突破。《木建筑》第2辑"大跨度木结构建筑专辑"就是收集了全球最新的、涵盖了各种建筑类型的设计案例集锦。

在钢铁和混凝土结构材料并未普及之前，世界上绝大多数的大跨度建筑都是天然的木竹结构。遍布世界各地的木结构桥梁（欧洲的德国、瑞士和奥地利，亚洲的中国，北美洲的美国和加拿大等国家遗存了大量的大跨度木构廊桥），从几米到几十米的拱架结构或者桁架结构，19世纪到20世纪北美洲的木构桥梁跨度竟然可以达到近百米；欧洲中世纪以来遗存的教堂、宫殿的桁架体系屋顶，工业革命以来的火车站大厅的木拱架等都是超过20米以上的大跨度木建筑；在亚洲，中国、日本和越南的佛殿和宫殿中也保存有超过10米以上的木结构建筑；在亚洲盛产竹子的国家，包括中国南部和越南，用原竹结构而成的大跨度建筑也时常见到，它们可以是宽敞的大厅和凉亭，也可以是堆放货物或煤炭的货栈。长时间大量运用天然有机材料作为建筑的结构材料，促使了以中国为代表的东亚国家逐步形成了"道法自然"的生态观和可持续发展的理念。

一百年前，欧洲人发明了现代胶合木为代表的工程木材料，使得木竹结构材料从传统的原木走向了胶合木为代表的现代木结构材料。有了现代结构科学和现代材料学的助力，木结构建筑体系可以超越传统时代的技术能力，让建筑跨越更大的空间。近十来年来涌现出一批大跨度木网壳结构的体育馆、仓库和温室项目，跨度都超过了100米。比如，2003年在美国塔科马市Tacoma建造的塔科马穹顶，木穹顶结构直径为162米，穹顶顶部高度达45.8米；2015年在意大利建成的Brindisi胶合木穹顶，该木穹顶结构直径约为143米，穹顶最高点高度可达49米；瑞士Haring木结构建筑公司在近10年建造的Saldome2木穹顶结构，可存储近100000立方米的盐，其直径约为120米，高度为32米。在亚洲，率先进入现代大跨度木结构建造的是日本，1997年他们就在秋田县大馆市建造了闻名世界的大馆树海体育馆，该体育馆平面呈椭圆形，由伊东丰雄建筑设计事务所与竹中工务店联合设计。该场馆平面沿长轴方向长度约为178米，短轴方向长度约为157米，穹顶高度约为52米，屋面木结构网壳体系主要采用秋田杉胶合木构件制成。最近，中国山西省太原市植物园内也建成了三个胶合木网壳结构的穹顶作为植物温室，最大的一个网壳平面尺寸直径达到88米，穹顶最高点达到29米。2011年，中国贵州省毕节市在百里杜鹃风景区也建成了一个大跨度胶合木网壳结构的多功能厅，其平面呈圆形，直径也达到了53米。江苏省的南京工业大学近年来一直致力于大跨度胶合木结构步行桥的设计与研究，在2011年他们建成了75米跨度的苏州胥虹桥，在2018年又建成了99米跨度的山东省滨州木拱桥。

实际上，现代木结构建筑体系跨越空间的潜力并未完全释放。我们期待，在未来的某一天，无论在中国，或是在亚洲，还是在欧洲和北美洲，乃至世界各地，我们都能很容易地进入新进崛起的一幢幢大跨度的木结构建筑中，它们或是机场大厅，或是体育馆，也可能是展览馆或者是火车站，也可以是一座步行桥，或者就是植物园中的温室，我们置身于其中，静静地去体验和欣赏它们的空间和展示的美……

Since the Industrial Revolution spread in Europe in the middle of the 18th century, with the gradual development of steel and cement production, large-span public buildings and transportation buildings have gradually been built by steel and concrete. For a period of time, the quality and quantity of steel produced has almost become the standard for measuring the level of industrialization to the country. To the construction industry, the way of steel wrapped in cement has become the only choice for constructing the main structure of the buildings. Even today, our construction industry is still unable to completely get rid of using material of the steel and cement. Although there is an irreconcilable contradiction between the production of steel and cement and the sustainable development of the human living environment, based on the current development level of science and technology and the basis of manufacturing, it seems still unrealistic to completely eliminate these two materials. Therefore, in order for us living on the planet to be healthier and longer-lived, the construction industry has to face the challenge: environmentally friendly and sustainable building materials must be used more for construction. The modern timber and bamboo structure construction system is exactly a solution explored in facing the increasing environmental crisis.

For a long time, the goal of building structure development has been developed within the two dimensions; one is height (designers and builders always hope that the buildings can be built taller and more upright and eye-catching, at the same time, saving precious land resources); the other is the span (designers and builders always hope that the span of the building will become larger, which can create a wider interior space and a magnificent space effect. The main structural materials that made the building taller and larger-spanned are still steel and reinforced concrete which are not surprising to know. Until now, under the promotion of science and technology and concepts of environmental protection, the modern timber and bamboo structure system has also joined to fight for the championship. It is gratifying that during the last ten years, the new sustainable structural material has not disappointed us. It has continuously refreshed the records of modern timber structures in height. Based on the traditional timber structure, new breakthroughs have also been made in span. The 2nd series of the book Mù Architecture (Large-span Timber Structure Architecture) is a collection of the latest design cases which covered all types of buildings around the world.

Before steel and concrete were not popularized, most large-span buildings in the world were built by timber and bamboo structure. There are many large-span timber bridges remained in the European countries, such as Germany, Switzerland, and Austria, in Asia, such as China, and North America, the United States and Canada. The span of the arch or truss structure of these timber bridges ranged from a few meters to tens of meters. The span of timber bridges built in North America from the 19th century to the twentieth century can reach nearly 100 meters. The churches and palaces with truss structure roof constructed since the European Middle Ages and the railway station hall with large timber arch built after the industrial revolution are the large-span timber architecture in more than 20 meters. In Asia, Buddhist temples and palaces built in China, Japan, and Vietnam, also have timber structure architecture with the span in more than 10 meters. In the Asian countries which are rich in bamboo, including south of China and Vietnam, large-span buildings built by raw bamboo structure are also commonly seen. They can be wide halls and pavilions, or warehouses that store goods or coal. Largely adopting natural organic materials as structural materials for long periods has motivated East Asian countries represented by China to gradually form an ecological view of following nature rules and the concept of sustainable development.

A hundred years ago, the Europeans invented the engineered timber represented by modern glulam. The material made for the wood and bamboo structure has changed from the traditional log to the modern material represented by glulam. With the help of modern structural and material science, the technical capability of the new timber structure construction system has surpassed the conventional system, making the building to be a larger space in span. In the past ten years, a number of large-span sports stadiums, warehouses and greenhouse projects with timber shell structure have emerged. The span is over 100 meters. For example, the Tacoma dome was built in Tacoma, USA in 2003. The diameter of the timber dome structure is 162m, and the height of the dome is 45.8m. The Brindisi glulam dome was built in Italy in 2015. The diameter of the timber dome structure is about 143m, and the highest point of the dome has reached 49m in height. The Saldome2 timber dome structure was built by Haring & Co.AG in Swiss in the past 10 years. It can store nearly 100,000 cubic meters of salt, and its diameter is about 120m and its height is 32m. In Asia, Japan was the first country to enter the field of the modern large-span timber structure construction. In 1997, the world-famous Odate Jukai Dome was built in Odate. The plan of stadium is oval and co-designed by Toyo Ito & Associates, Architects and Takenaka Corporation. The length of the stadium's plan is about 178m along the long axis, about 157m along the short axis, and the dome height is about 52m. The timber net-shell structure of the roof is mainly made of Akita cedar glulam. Recently, three domes of glulam reticulated shell structure, as a plant greenhouse, have been built in the Botanical Garden of Taiyuan City, Shanxi Province in China. The largest dome has a diameter of 88 meters in plan and the highest point of the dome reaches 29 meters. In 2011, a multi-span multi-purpose hall with a long-span glulam net shell structure was built in Baili Dujuan ("Hundred Miles of Rhododendrons") Scenic Area in Bijie City, Guizhou Province, China. Its plan is circular and its diameter has reached 53 meters. Nanjing Tech University in Jiangsu Province has been dedicated to the design and research of large-span glulam footbridges in recent years. The Xuhong Bridge with 75m span was built in 2011. The Binzhou timber arch bridge with 99m span was built in Shandong Province in 2018.

In fact, the potential for spanning space of modern timber structure is more than that. We look forward to be able to easily enter the newly-rising large-span timber structure architecture, on one day in the future, whether in China or in Asia, or in Europe and North America, or even around the world. They might be the airport hall, gymnasium, exhibition hall or railway station, or a footbridge, or a greenhouse in a botanical garden. We quietly experienced in side of it and appreciated the space and the beauty displayed.

上海交通大学设计学院木建筑研究与设计中心主任
建筑学系教授、博士生导师
The Center for Timber Architecture Research & Design
Professor and Doctoral Supervisor of Architecture
School of Design, Shanghai Jiao Tong University

2019 年 11 月
November 2019

序
Preface

欣闻上海交通大学设计学院刘杰主编的《木建筑》第2辑即将面世，很高兴为此写几句话。

众所周知，中国当前已经进入了一个生态文明建设的新时代，中国政府向世界庄严承诺：中国计划2030年二氧化碳排放达到峰值且将尽早达峰。到2030年单位GDP二氧化碳排放比2005年下降60%到65%。但是，目前中国在很多方面还面临比较严峻的挑战。以建筑业材料生产和使用为例，我国粗钢产量已占全球近一半；二氧化碳排放约13亿吨，约占我国总排放12%。我国水泥产量占全球一半以上，水泥工业年消耗标煤约2亿多吨，占建材行业能源消耗总量的75%。我国建筑业碳排放所占比重达到了50%。由于水泥、钢铁等材料均产自不可再生资源，其生产能耗大、污染比较重。相对来说，作为一种重要的环保建材，木材的生产和在建筑中的使用还存在很大的发展空间。

从环保角度看，木材属于可再生材料，木结构对环境的污染比钢结构、混凝土结构少很多。木结构比混凝土结构节能8%~16%；每立方米木材将储存0.9吨当量的二氧化碳，同时在木材生产制造阶段还可减少1.1吨碳排量。党和国家高度重视建筑业木结构的发展。中共中央国务院关于进一步加强城市规划建设管理工作的若干意见（中发〔2016〕6号）指出："发展新型建造方式：积极稳妥推广钢结构建筑。在具备条件的地方，倡导发展现代木结构建筑。"工信部则发文指出：促进城镇木结构建筑应用，推动木结构建筑在政府投资的学校、幼托、敬老院、园林景观等低层新建公共建筑，以及城镇平改坡中使用。推进多层木—钢、木—混凝土混合结构建筑，在以木结构建筑为特色的地区、旅游度假区重点推广木结构建筑。在经济发达地区的农村自建住宅、新农村居民点建设中重点推进木结构农房建设。笔者前两年也在第十届江苏省园艺博览会主展馆建筑中，与葛明教授和徐静高级建筑师等合作，联合南京工业大学木结构研究所刘伟庆教授和陆伟东教授团队，建筑主体运用了木结构，营造了独特的建筑形象和空间环境，获得了良好的效果。此外，我们也在南京江宁钱家渡的新农村建设中采用了木结构。

多年来，上海交通大学设计学院刘杰教授团队在传统木结构和现代木结构的结合方面深耕多年，取得重要的研究和工程实践成果，在业界享有盛誉。本期《木建筑》第2辑主题为"大跨度木结构建筑"专辑，收录了中国、美国、英国、加拿大、法国、挪威、荷兰、奥地利、澳大利亚等多个国家的具有代表性的大跨度木结构建筑工程。建筑类型涵盖体育建筑、文教建筑、交通建筑、会展建筑、商业办公建筑、园林建筑、纪念性建筑和工业建筑等。其中包括由扎哈·哈迪德建筑事务所设计的英国森林绿色漫游者体育馆，获得多项国际设计大奖的加拿大列治文椭圆冬奥速滑馆，在业内受到诸多好评的重庆龙湖两江长滩原麓社区中心等具有设计和技术创新的典型木结构建筑项目。本辑为专业和非专业领域内的读者深入浅出地介绍大跨度木结构领域的最新技术和应用成果，例如胶合木结构体系和标准化木构件连接方式等方面，展现大跨度木结构独特的美，同时，分析木材特性如何与功能完美结合的实际案例和具体做法。期盼为发展中国的大跨度木结构建筑设计和技术领域，提供一定的借鉴。

I am glad that the second series of *Mù Architecture* edited by LIU Jie from the School of Design of Shanghai Jiao Tong University is about to be released, and I am very happy to write a few words for this.

As we all know, China has entered a new era of construction of ecological civilization, and the Chinese government has solemnly promised the world that China plans to reach a peak of carbon dioxide emissions by 2030 and will reach its peak as soon as possible. By 2030, carbon dioxide emissions per unit of GDP will be 60%~65% lower than in 2005. However, China still faces severe challenges in many aspects. Taking the production and use of construction materials as an example, China's crude steel output accounts for nearly half of the world in total; carbon dioxide emissions are about 1.3 billion tons, accounting for about 12% of the total emissions in China. China's cement output accounts for more than half of the world in total. The cement industry consumes about 200 million tons of standard coal annually, accounting for 75% of the total energy consumption of the construction material industry. Carbon emissions of construction industry in China account for 50% in total. Because cement, steel and other materials are produced from non-renewable resources, their energy consumption for production is large and pollution is relatively heavy. Relatively speaking, as an important environmental friendly building material, there is still great development space for wood in the production and construction.

From the perspective of environmental protection, wood is a renewable material, and the pollution of timber structure to the environment is much less than that of steel and concrete structure. Timber structure is 8%~16% less energy consumption than concrete structure; each cubic meter of wood will store 0.9 tons of equivalent carbon dioxide, and at the same time, it can reduce 1.1 tons of carbon emissions during the its manufacturing stage. The Party and the state attach great importance to the development of timber structure in the construction industry. CPC Central Committee and State Council proposed several opinions on further strengthening the management of urban planning and construction (ZF〔2016〕No. 6) which was pointed out that developing new construction methods and promoting steel structure buildings steadily, and advocating the development of modern timber buildings, where conditions permit. The Ministry of Industry and Information Technology of People's Republic of China pointed out that the application of timber architecture in the urban area is being promoted, and it also promote to adopt timber structure for new low-rise public buildings, such as government-invested schools, nurseries, homes for the elderly, garden landscapes, as well as in the project of urban apartment roof renovation. It mainly promotes to build multi-story timber-steel, timber-concrete hybrid structure buildings, and promotes timber structure buildings in wood-featured areas and tourist resorts. The timber structure is also promoted to apply in rural self-built houses of economically developed areas and new rural residential areas. In the first two years, the author cooperated with Professor GE Ming and Senior Architect XU Jing to design the project of the main exhibition hall in the 10th Horticultural Exposition of Jiangsu Province. The project is also in conjunction with the team of Professor Liu Weiqing and Professor Lu Weidong from the Institute of Timber Structure of Nanjing Tech University. The main structure of the building is the timber structure, which has created a unique architectural image and environment, and it has achieved good results. In addition, the timber structure is also adopted in the construction of a new countryside in Qianjiadu village of Jiangning, Nanjing.

For many years, the team of Professor LIU Jie from the School of Design of Shanghai Jiao Tong University has been focused on the research of traditional and modern timber architecture for many years, and has got important research and practice achievements, enjoying a high reputation in this field. The theme of this second series of *Mù Architecture* is the album of the large-span timber architecture. It includes representatively large-span timber construction projects from China, the United States, the United Kingdom, Canada, France, Norway, the Netherlands, Austria, and Australia. The types of buildings cover sports buildings, cultural and educational buildings, transportation buildings, exhibition buildings, commercial office buildings, landscape buildings, monumental buildings, and industrial buildings. It includes many typical cases with design and technological innovation, such as the British Forest Green Rover Stadium designed by Zaha Hadid Architects, the Richmond Olympic Oval in Canada, which has won many international design awards, and Yuanlu Community Center in Chongqing, which has received many praises in the industry. This album provides readers in the professional and non-professional fields with an easy-to-understand introduction to the latest technology and application achievements in the field of large-span timber structure. For example, glulam structural systems and standardized connection methods of timber components. It also shows the unique beauty of large-span timber structure. At the same time, it analyzes practical cases and specific methods of how characteristics of timber material and its construction functions can be perfectly combined. It will be able to a good reference for developing the design and technology of China's large-span wooden buildings.

2019 年 11 月
November 2019

王建国：中国工程院院士，东南大学教授，东南大学城市设计研究中心主任
WANG Jianguo: Academician of Chinese Academy of Engineering (CAE), Professor of Southeast University, Director of Urban Design Research Center of Southeast University

目录
Contents

主编的话	v	Editor's notes
序	viii	Preface

体育建筑
Sports Building

英国费德曼学校新泳池	001	New Swimming Pool in Freemen's School, UK
英国森林绿色漫游者体育馆	013	Forest Green Rover Stadium, UK
加拿大列治文奥林匹克椭圆速滑馆	018	Richmond Olympic Oval, Canada
加拿大观景山庄水上中心	034	Grandview Heights Aquatic Centre, Canada
日本东京新国立体育馆	046	New National Stadium in Tokyo, Japan
长春市全民健身活动中心游泳馆	051	National Fitness Center in Changchun Swimming Pool, China

文教建筑
Cultural & Educational Building

澳大利亚班吉尔广场	054	Bunjil Place in Casey, Australia
重庆龙湖两江长滩原麓社区中心	063	Yuanlu Community Center in Chongqing
加拿大太平洋自闭症家庭中心	081	Pacific Autism Family Centre, Canada
加拿大夸扣特尔瓦加鲁斯学校	087	Kwakiutl Wagalus School, Canada
加拿大北不列颠哥伦比亚大学的木材创新研究实验室	090	Wood Innovation Research Lab, University of Northern B.C., Canada

交通建筑
Transportation Building

| 菲律宾麦克坦—宿雾国际机场 | 096 | Mactan-Cebu International Airport, Philippines |
| 挪威奥斯陆机场第二航站楼 | 105 | Oslo Airport, Terminal 2 in Oslo, Norway |

会展建筑
Exhibition Building

上海西郊宾馆意境园多功能厅	116	Yijingyuan Multi-function Hall in Xijiao State Guest Hotel, Shanghai
上海西岸人工智能峰会B馆建造实践	133	West Bund World AI Conference Venue B, Shanghai
天津欢乐谷演艺中心	142	Art & Performance Center of Happy Valley, Tianjin
贵州紫云自治县格凸河攀岩基地观赛广场	148	Spectator Square of the Climbing base, Getu River, Ziyun Autonomous County, Guizhou
江苏省委会议厅	154	A Conference Hall in Jiangsu Province
云南弥勒太平湖森林小镇国际木屋会议中心	168	Taiping Lake International Conference Center in Forest Town of Mile City, Yunnan

商业办公建筑
Commercial & Office Building

加拿大素里城市中心	173	Surrey Central City, Canada
美国竖屏沃敏斯特总部办公楼	180	Vertical Screen Warminster Campus, USA
美国华盛顿水果生产公司总部办公楼	189	Washington Fruit & Produce Co. Headquarters, USA

园林建筑
Landscape Building

| 荷兰库肯霍夫公园中心 | 199 | Keukenhof Flower Garden Center, Netherlands |
| 太原植物园 | 206 | Taiyuan Botanical Garden, China |

纪念性建筑
Commemorative Building

挪威阿尔嘉德的新木质教堂　210　Ålgård's new wooden church, Norway

工业建筑
Industrial Building

加拿大"结构工坊"工厂	219	Structurecraft Manufacturing Facility, Canada
奥地利菲沙门德物流中心	220	Logistics Center in Fischamend, Austria
法国喷气式飞机库	233	Garage for the Jetset, France

专题
Specialty

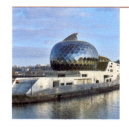

法国和瑞士的创意屋顶　240　Creative Roofs in France and Switzerland

工程名录
Engieering Directory

世界大跨度木结构建筑一览表（2000—2020）　262　List of World Large Span Wood Structural Buildings from 2000 to 2020

英国费德曼学校新泳池
New Swimming Pool in Freemen's School, UK

作者：Erik Bredhe
翻译：蒋音成
校对：沈姗姗
摄影：Jack Hobhouse, Hawkins/Brown

The new pool at Freemen's School in Surrey was the second phase in the modernisation of the school. Phases three and four include the refurbishment of the Grade II* Listed Main House and landscape improvements. The pool was designed by British architects Hawkins/Brown, who are based in London and Manchester. Founded in 1988, the practice now comprises over 250 architects, interior designers, city planners and researchers.

Client: City of London Freemen's School
Constructor: Gilbert-Ash
Cosy: EUR 9,3 million
Size: 1750 m²

Wood in a historical setting. Large, light glulam vaulting makes the swimming pool at Freemen's School warm, inviting – and stunning. A building expected to deliver on form and function has taken shape in old oak woodland in the school grounds.

The old swimming pool at freemen's school rose high above the rolling landscape of Surrey in the UK. Since it was built in the 1970s the pool, which was also used for diving competitions, had become something of a landmark in the area. Like many other parts of the school, the building sat in an area of historical importance. It was therefore a particular shame when the old pool burned down one rainy January night in 2014.

Six years earlier, British architectural practice Hawkins/Brown had been commissioned to redesign the whole of the campus that makes up the City of London Freemen's School, a boarding school in Surrey, south of London. By 2014, they had just completed phase one of four. The swimming pool – which was actually due to be built in the final phase – was immediately given a higher priority.

"Other buildings were put on hold and we began work on the swimming pool earlier than planned. But when we got down to the drawings, we felt we wanted to make a number of changes. We were set to redevelop the old swimming pool, but then we decided to place the pool in a location that felt more natural." says Adam Cossey of Hawkins/Brown.

位于英国素里的费德曼学校新泳池是该学校现代化改造进程中的第二期工程。其第三和第四期工程包括翻新二年级学生教学主楼和景观改造。这个泳池是由英国设计师 Hawkins/Brown 设计的。他们的办公室位于伦敦和曼彻斯特。公司建立于 1988 年,目前有超过 250 名建筑师、室内设计师、城市规划设计师和研究人员。

客户:伦敦费德曼城市学校
施工方:Gilbert-Ash
花费:9.3 亿欧元
规模:1750 平方米

木结构建筑形式的历史渊源。巨大的轻型胶合木拱顶让费德曼学校的游泳池焕发出温暖、诱人和惊艳的光泽。这栋建筑的形式和功能外形取自于学校场地中的那片老橡木林。

费德曼学校的老游泳池曾高耸于英国素里市的起伏景观区中。这座游泳池始建于 19 世纪 70 年代,被用作潜水的比赛场地,并成为当地的地标性建筑。就像学校的其他建筑一样,这栋建筑也坐落在重要历史区域内。但非常遗憾的是,这座老游泳池在 2014 年 1 月的雨夜中被一场大火烧毁了。

6 年前,英国建筑事务所 Hawkins/Brown 被委任重新设计费德曼学校的整个校区。费德曼城市学校是一所位于伦敦南部的寄宿学校。截至 2014 年,他们完成了 4 期工程中的第一期。此游泳池本来安排在最后一期建造,但是它却被给予了更高的优先建造权。

Hawkins/Brown 的 Adam Cossey 说:"其他建筑的设计工作被暂停了。我们比原计划更早地开始了游泳池项目的设计工作。但是当我们开始画图纸的时候,我们发现设计需要做出较多的改动。我们本来被委派原地重建老游泳池,但是我们决定将游泳池放在一个让人觉得更自然的地方。"

The pool was relocated from the west of the campus to the east, adjacent to the rest of the sports facilities such as the rugby pitch and cricket field, football pitch and indoor tennis courts. Something else nearby that caught the architects' attention was an area of beautiful old oak woodland. They began to dream about getting to build here, inside the woods. However, that brought major challenges, not least in getting permission to use the land. The landscape, like the school's stately Main House and other parts of the site, is listed for its historical value. Following long negotiations, the architects eventually obtained permission to build in this location. There was one condition, though – they had to create an attractive building that worked well with the other buildings and accentuated rather than overwhelmed the woodland.

"The surrounding greenery became a major source of inspiration for the pool. We already knew we wanted to build in wood, but that choice ended up being even more obvious. We wanted the woodland to be reflected in the interior, so we created an exposed glulam design. You look straight out into the woods through the large windows, but there's also wood in the walls and ceiling. It gives a sense of swimming among the trees."

　　这座游泳池被从校园的西面移到了东面，毗邻其他体育设施，例如橄榄球和板球场、足球场和室内网球场。除此之外，场地内另一些吸引建筑师的因素，就是那片美丽的老橡树林。他们开始想象把游泳池建在林子里的样子。但是，这样的想法给他们带来了巨大挑战，尤其是得到那块地的建设许可。这块景观区域，跟学校的主楼和其他部分一样，被列入历史保护名单。经过漫长的协商后，建筑师最终获得了这块区域的建设许可。但有一个条件就是他们必须设计一栋独具魅力的建筑，同时，这栋建筑要与周围的建筑物相契合，衬托而不是弱化那片林地。

　　"周围的绿色景观成为这座游泳池的巨大灵感来源。想把建筑建在林子中的想法由来已久，但是这个选择在最后变得更明确。我们希望建筑的室内能够映射出这片林地，所以我们使用了外露式的胶合木设计。你不仅可以从巨大的窗子里看到林地，也可以从墙和屋顶中看到林子。这样的设计营造了一种在树丛间游泳的感觉。"

总平面图 Site Plan

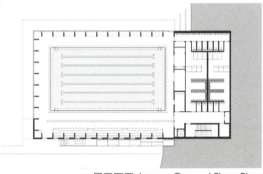

一层平面图 Lower Ground Floor Plan

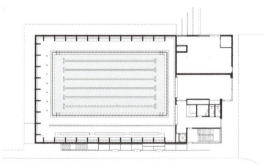

局部二层平面图 Upper Ground Floor Plan

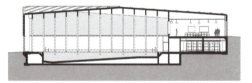

剖面图 Section

The pool's structure comprises large portal frames of glulam, with panels of CLT forming the building's walls and ceiling. Looking up at the vaulted frame, it becomes apparent that some of the beams are jointed and extended. The glulam components are fixed together with dowel joints and inset plates. There is a natural explanation for this approach – the narrow roads in the British countryside made it impossible to transport huge components.

The whole of the prefabricated structure was assembled in a little over two weeks. The pool's inspiring design is as simple as it is beautiful. Hawkins/Brown took the boxy, functional shape that swimming pools often have, but played with the concept. Pushing down two of the box's diagonally opposite corners and pushing up the other two created fascinating angles and spaces. The fact that part of the building is set into a gentle slope further distorts the perspectives. Adam Cossey explains that although the structure might look advanced, it is actually quite simple.

"One half is in fact just a mirror-image of the other, which made production of the glulam frames easier. Raising and lowering the pool's corners also had a practical function. The roof drops slightly where the site slopes, so the building nestles into the landscape. Where the building rises up, on one side we've placed the entrance and on the other side large windows facing the woods."

　　游泳池的结构采用了巨大的胶合木门式框架结构，墙和屋顶采用了CLT板。注意看这个拱形框架，一些梁连接在一起并延伸出去。这些胶合木构件由销钉和插板固定到了一起。这样的方式有一个很客观的原因：英国的乡村小路没有办法运输巨大的部件。

　　整个装配式结构在两周多一点的时间内组装完成。这个设计既简单又好看。Hawkins/Brown选择了四方的具有功能性的常见形态（泳池通常是方形的），并且在此基础上进行创新。将盒子的两个斜对角推进去，再将另两个斜对角拉出来，通过这个手法创造出迷人的角度和空间。部分建筑设计为一个缓坡；随后形体扭曲改变其透视角度。Adam Cossey解释到，"虽然这座建筑看起来很先进，但是它其实很简单"。

　　"建筑一半与另一半是对称的；这使得生产胶合木框架更容易。拉高和推低游泳池的空间一角也有着现实作用。屋顶随着地形的坡度而倾斜，因此这栋建筑也半隐半现于景观之中。在建筑屋顶升起的区域，一侧设置了建筑的主入口，另一侧为面向树林的大落地玻璃。"

All the interior wood is finished with a white stain, primarily to give the pool a light feel, although coupled with the large expanse of glazing, it also reduces the need for lighting. In addition, the stain contains a treatment that fireproofs the building. It was important, however, that you could still see the grain of the wood through the pale white finish.

"We wanted to make the most of wood's warm, comforting properties in the swimming pool. It was one of the reasons why we chose to keep all the wood exposed. It's also nice for the students who swim here in the day to see the structure. When you see it, you can understand how the building is put together and so it becomes educational and instructive."

Swimming pools tend to be tricky to handle acoustically, with sound readily bouncing around the large hall, making sound levels problematic. But the excellent sound-damping capacity of the wood makes this much less of an issue in the Freemen's School pool. They have also worked on the acoustics by placing sound-absorbent wooden panels vertically along the walls in the actual pool area. In addition to its acoustic properties, there are many other reasons why wood is so well suited to swimming pools. Wood offers good durability in large buildings with high humidity levels. While steel, for example, corrodes, wood handles the moisture better and also has thermal insulation properties. However, the architects were forced to moisture-proof the glulam frames structurally nearest the floor.

虽然外墙采用了大面积的玻璃，所有室内木材都被处理成了白色，以帮助游泳池营造轻松的氛围，同时也减少了照明需求。此外，涂料也具备防火功能。重要的是人们依旧可以通过白色涂料看到木材本身的纹理。

"在这座泳池建筑中，我们想要尽量发挥木材温暖和舒适的功效。这是我们选择暴露所有木结构的原因之一。当学生们游泳时，可以看见这些木构件，也是很美好的事。当人们看到木构件的时候，他们会明白一栋建筑是怎么建构而成的。这样子的话，这座游泳池就具有教育和指导意义。"

设计游泳池的隔声方面本身很需要技巧：声音会在大的空间里来回反射，这让音量控制变得困难。但是木材本身极好的消声特性让这个问题在费德曼学校的游泳池项目中变得很小。建筑师们通过在游泳池周边的墙面上铺设竖向吸声板，来解决声效问题。除了木材的吸声特性，还有很多其他原因使木材成为这个项目的最好选择。木材在潮湿的大型建筑中具有较好的耐久性。举个例子，钢材会生锈，但是木材在高湿度的环境里表现得更好，也具有较好的保温性能。建筑师被要求在离地面最近的地方采用做防潮处理的胶合木框架。

"We've placed the portal frames on a kind of steel foot that was designed by Wiehag, who supplied the glulam structure. These were specially developed to keep the glulam structure from standing in water while also supporting the load."

In addition to the 25 m pool with six lanes, the building also houses changing rooms, a combined classroom and function room and a reception. Wood also runs through these areas, with walls of CLT and particleboard ceilings. Externally, the architects have used larch at the entrance, a material that will silver attractively over time. The vertical strips of larch are a reference to the first phase of the redevelopment of Freemen's School – the boarding house – which ties together the design language of the buildings. The rest of the exterior comprises large windows, with zinc standing-seam cladding on the walls and roof. The zinc cladding was important in the wider context, as the school's Main House has similar roofing. In addition to sitting comfortably amongst the surrounding woodland and lawns, the new swimming pool also had to work well with the Grade II* listed Main House.

The students, the client and the architects are all pleased with the end result. Adam Cossey is particularly proud of how they have managed to keep the pool stylish and minimalist. All the functions have been dealt with ingeniously and built into the design.

The lighting has been concealed in fittings along the top edge of the windows. The ventilation is hidden in the floor adjacent to the windows, blowing out air to prevent condensation, a little like with a car windscreen.

"Swimming pools tend to be quite busy, with a lot of people and numerous functions to accommodate. I think we've managed to create something new at Freemen's School. A scaled-back, functional and aesthetically pleasing building that welcomes visitors with a warm embrace. And all the time there's the old oak woodland as a tranquil and reassuring backdrop."

"我们采用了由 Wiehag 设计和供应的结构构件，此结构形式是将门式框架胶合木构件放置在钢基础之上。这种方式既满足胶合木结构能立在水里，同时能够承受屋顶荷载。"

除了拥有 6 条泳道、25 米长的游泳池以外，这座建筑也包含了更衣间、多功能教室和前台。木材也被用在这些区域（CLT 的墙和刨花板做的吊顶）。对于建筑外立面，建筑师们在入口处采用了落叶松木，它将会随着时间的推移变得更闪亮迷人。竖向的落叶松木条装饰是费德曼学校改造计划第一期工程的参照，即寄宿大楼，此装饰木条跟建筑的设计语言是有联系的。余下的建筑外立面，包括了巨大的窗子和镀锌板覆盖的外墙面和屋面。由于学校的主建筑也采用了同样的屋顶形式，此处的镀锌外挂板从广义上来说，是很重要的。新游泳馆除了坐落于舒服的林地和草地之中，它和二年级主楼也非常协调。

学生、客户、建筑师都对最终的结果很满意。Adam Cossey 尤其自豪于他们让这个游泳池带着极简主义和时尚感。他们创造性地设计了所有功能，也把功能很好地融入建筑中。

照明设备隐藏于窗户上边缘的配件中。通风设备被藏在窗户旁边的地板里；这些设备吹出空气以防止形成冷凝，有一点像汽车的挡风玻璃。

"游泳池由于使用者和多样的内部功能而变得十分忙碌。我认为我们为费德曼学校带来了一些创新之物，一栋小规模的、功能性齐全、美观的建筑，正以它温暖的拥抱迎接着游客。同时，有一片老橡木林，为其提供安静和安心的陪衬景色。"

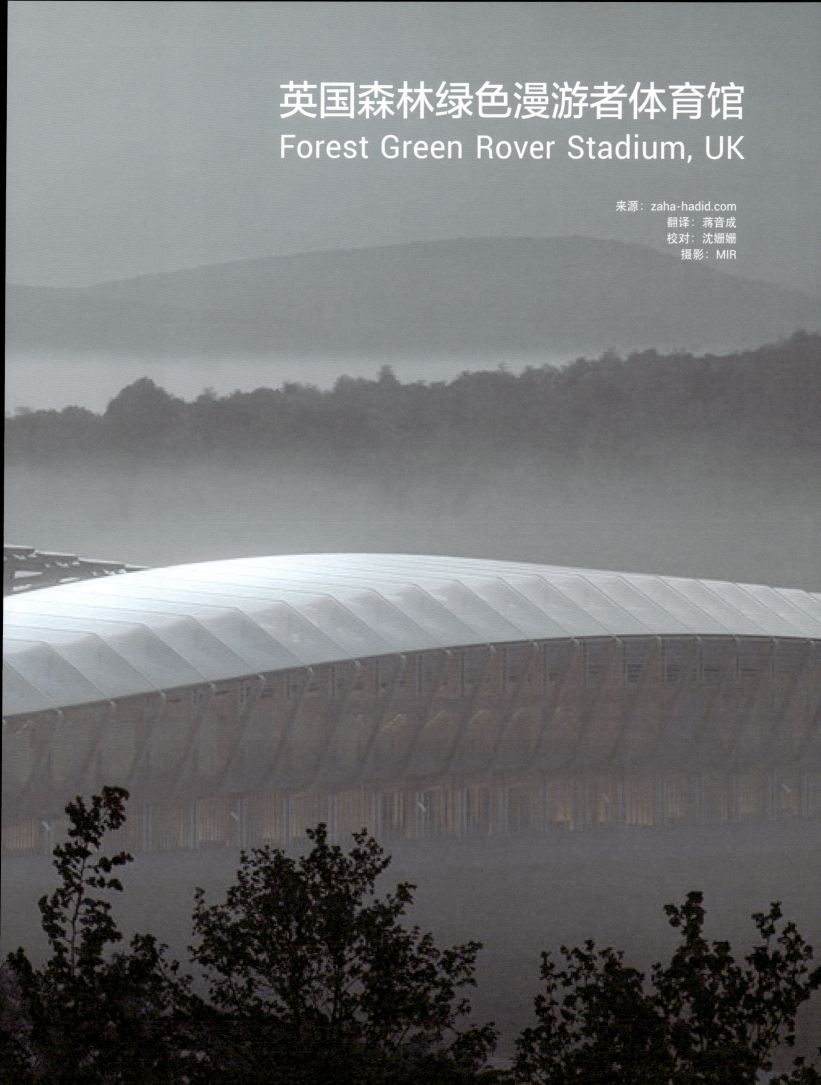

英国森林绿色漫游者体育馆
Forest Green Rover Stadium, UK

来源:zaha-hadid.com
翻译:蒋音成
校对:沈姗姗
摄影:MIR

The British Football Club Forest Green Rovers is getting a new stadium. The winner of the major international competition was Zaha Hadid Architects, whose entry was a stadium built almost entirely of wood. In phase 1, the stadium will have capacity for 5000 spectators, with the option of expanding this to 10000 seats when the club reaches the higher leagues.

英国森林绿色漫游者足球俱乐部正在筹建一个新的体育馆。这个项目中标的赢家是扎哈·哈迪德建筑事务所；他们提交的方案是一个几乎完全用木材建构的体育馆。在第一阶段，这座体育馆可以容纳5000名观众。在俱乐部达到更高级别的联盟规模时，俱乐部可以扩大体育馆的容量至10000个座位。

Almost all of the stadium, from the structural frame and the roof brackets to the screen that clads the exterior, will be sourced from sustainably managed forests. The design of the load-bearing structure allows even the stands and the floor components to be made of wood. The roof will be clad in a thin transparent membrane that allows enough light through to keep the grass pitch healthy. The club calculates that their new arena will have the lowest carbon footprint of any football stadium in the world.

体育馆从结构框架、屋顶托架到覆盖外墙的支撑件,其使用的几乎所有材料都来自于可持续发展式管理的森林。支撑荷载的结构设计使得看台和楼面构件也可使用木材来建造。体育馆的屋顶上覆盖了一层透明的防水材料,使得阳光可以充分射入,来确保草地的健康生长。俱乐部对体育馆进行的计算表明,这座新体育馆的碳排放量比世界上任何的体育馆都要低。

加拿大列治文奥林匹克椭圆速滑馆
Richmond Olympic Oval, Canada

来源：Naturally Wood
翻译：高晓萌
校对：沈姗姗
摄影：Craig Carmichael, KK Law, Stephanie Tracey, Martin Tessler

Total Building Area: 47000 m²
Client: City of Richmond
Architect: Cannon Design
Roof structural engineer: Fast + Epp Structural Engineers
Project management: MHPM Project Managers Inc.
Construction manager: Dominion Fairmile Construction Ltd.
Building renderings: Cannon Design
Woodwave design-builder: StructureCraft Builders Inc.

Awards:
2015   All Time Award, International Association of Sports and Leisure Facilities and the International Olympic Committee
2010   Facility of Merit, Athletic Business Magazine
2010   Sports Building of the Year, ArchDaily Building of the Year Awards
2010   Green Good Design Award, Chicago Athenaeum: Museum of Architecture and Design
2010   Award of Merit, ARIDO
2010   Citation, Society of American Registered Architects, (SARA) National
2009   Best of Canada Design Award Canadian Interiors, Best in Canada Design Awards
2009   North American Wood Designer Award, Canadian Wood Council
2009   Interior Design Honor Award, Inform Awards
2009   Award of Excellence for Innovation in Architecture, Royal Architectural Institute of Canada
2009   Excellence for Green Building Award, The Globe Foundation and the World Green Building Council
2009   Sustainability Star Award, Vancouver Organizing Committee
2009   Chairman's Trade Award, Gold, VRCA Awards of Excellence
2009   Chairman's Trade Award, Silver, VRCA Awards of Excellence
2009   President's Trade Award, Silver, VRCA Awards of Excellence
2009   Institutional Wood Design, Wood WORKS!

"The architectural design of the Richmond Olympic Oval emanates from several poetic images based in the cultural history of the site and the surrounding geography. For example, the articulation of the Oval roof evolved from the image of the Heron being a native bird in that community.

The roof has a gentle curve that peels off on the north side of the facility, emulating the wing of a heron with its individual feather tips extending beyond the base wood truss structure. This allows for the opening of the facility's interior to a view of the north shore mountains and the Fraser River at the North Plaza."

— Larry Podhora, Cannon Design

建筑面积：47000平方米
客户：列治文市
建筑设计：Cannon Design 建筑设计公司
屋顶结构工程设计：Fast + Epp 结构工程咨询公司
项目管理：MHPM 项目管理公司
施工经理：Dominion Fairmile 建筑有限公司
建筑渲染图：Cannon Design 建筑设计公司
"木浪"设计及建造：StructureCraft Builders 建筑设计公司

获奖：
2015年全时奖，国际运动休闲设施协会国际奥林匹克委员会
2010年优秀设施，运动商业杂志
2010年年度体育运动建筑，ArchDaily 年度建筑奖
2010年绿色优秀设计奖，芝加哥雅典娜建筑与设计博物馆
2010年优秀奖，安大略注册室内设计师协会
2010年引用奖，美国注册建筑师协会
2009年最佳加拿大室内设计奖，加拿大最佳设计奖
2009年北美木建筑设计师奖，加拿大木建筑协会
2009年室内设计荣誉奖，Inform 杂志奖
2009年杰出创新建筑奖，加拿大皇家建筑协会
2009年杰出绿色建筑奖，世界基金会和世界绿色建筑委员会
2009年可持续之星奖，温哥华组织委员会
2009年会长贸易金、银奖，温哥华建筑协会杰出奖
2009年主席金贸易银奖，温哥华建筑协会杰出奖
2009年木结构设计奖，Wood WORKS!

"列治文奥林匹克椭圆速滑馆的建筑设计灵感，来源于以当地文化史和周围地理环境为基础的诗歌意象。例如，椭圆屋顶的建筑形式是从一种当地土生土长的名叫苍鹭的鸟的外形演变而来的。

该屋顶是一条柔美的曲线（延展至建筑的北面），犹如鹭鸾的翅膀，它的独特的羽毛尾翼延伸出去，出挑于整个木桁架结构之外。这样的设计使得观众在建筑室内就可欣赏到北岸山脉风景和北边广场的菲沙河。"

—— Larry Podhora, Cannon Design 建筑设计公司

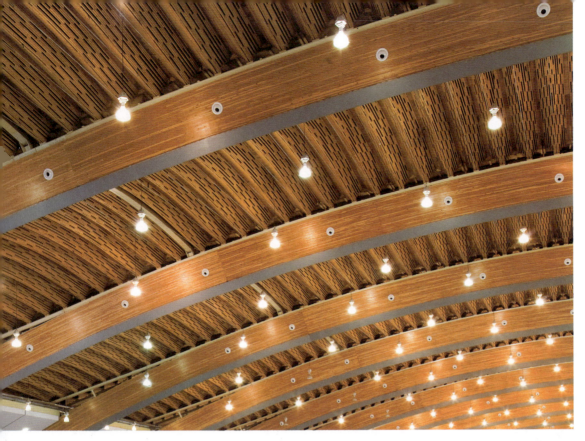

The Richmond Olympic Oval was a signature structure for the 2010 Olympic Winter Games and is a precedent-setting example of advanced wood engineering and design. The design of the wood roof and its application in a building of this size and significance marks the entry of British Columbia's wood design and fabrication industry onto the world stage.

The design concept of flow, flight and fusion was inspired by the water of the nearby Fraser River, the wild birds that inhabit its estuary and the careful meshing of forms—curved and linear—where city and nature meet.

The building is arranged on three levels: an underground parking garage; a ground-oriented entry, circulation, service and amenity level; and the breathtaking volume of the vaulted sports hall on the top level.

For the 2010 Olympic Winter Games, the Oval housed a 400-metre speed skating track with temporary capacity for approximately 8000 guests. After the Games, the facility was converted to multi-purpose sports use. The main sports hall has become an indoor activity area divided into three sections: ice, court, and track and field.

The ice section has two ice rinks. The court section is a combination hardwood and rubber surface playing area capable of hosting a wide variety of sports, while the track and field section has a rubberized turf surface that is home to an indoor running track and other sports. The space is convertible to different configurations that allow the facility to be used for a combination of ice and dry sports as demand warrants, including occasional reconfiguration for major short track and long track speed skating events.

One of the 2010 Olympic Winter Games' most prominent buildings and venue for the long track speed skating events, the Richmond Olympic Oval features a one-of-a-kind, all-wood roof structure.

The structure comprises composite wood-steel arches, which span approximately 330 ft (100 m), with a hollow triangular cross-section that conceals mechanical, electrical and plumbing services. Spanning the 42 ft (12.8 m) between arches are novel, prefabricated Wood Wave Structural Panels © consisting of ordinary 2 in × 4 in (38 mm × 89 mm) lumber arranged geometrically to optimize both structural and acoustic efficiency. The design is not only economical but it provides a striking aesthetic for this high-profile facility.

列治文奥林匹克椭圆速滑馆是2010年冬季奥运会标志性建筑，同时也是先进木结构工程设计的应用先例。木结构屋顶设计及其在如此规模的建筑物上的应用是具有重要意义的，标志着不列颠哥伦比亚省的木材设计和制造产业登上了世界舞台。

"流动、飞翔、融合"的设计灵感来源于场地附近的菲沙河波浪起伏的流水与栖息于河口野生的雀鸟，两者之间巧妙的契合形式——曲线与直线元素——如同城市与自然的融合。

该建筑共有三层：地下车库层；入口、流通、服务和设备设施层；以及顶层的巨型拱顶体育馆。

用于2010年冬季奥林匹克运动会的椭圆速滑馆，拥有400米的速滑道，最多可容纳约8000名观众。冬奥会结束后，该设施转变成适合各种运动的多用途运动馆。体育馆的主厅已转成室内活动区域，其被分成三部分：冰上运动场、球类运动场和田径运动场。

冰上运动场设有两个室内溜冰场。球类运动场采用了硬木和橡胶结合的地面，适用于举办各种体育活动；田径运动场的地面，铺了一层橡胶草皮，可用作室内跑道和其他运动场地。整个空间可转变为多种不同空间形态，以满足该建筑设施适用于各种冰上或地面运动的需求，包括临时改建场地用于短道速滑和长道速滑比赛。

作为2010年冬季奥运会最杰出的建筑之一和长道速滑比赛场馆，列治文奥林匹克椭圆速滑馆的最大特点是其举世无双的木结构屋顶。

该建筑结构由复杂的钢木混合拱形结构组成，跨度约为100米。其结构构件组装后呈空心三角形，中部空间内可隐藏机械、电气和管道设施。工厂预制的"木浪"结构板横跨于间距约为12.8米的曲梁之间，该"木浪"结构一般由38毫米×89毫米SPF规格材构成，通过几何设计使其兼顾了结构牢固和吸声效果。该设计不仅仅经济合理，同时给这一举世瞩目的设施增添了不少美感。

"To me the roof is the most spectacular part—from the inside it's like looking up at the stars."
    kristina Groves, 2008 world cup speed skating champion

The two lower floors of the building are cast-in-place concrete. Extending through the floors are massive, inclined concrete buttresses from which the great arched roof springs. The main arches, 47 ft (14.3 m) apart, are comprised of twinned glulam members held at an angle to one another by a steel truss. These arches conceal the building's mechanical and electrical services in their triangular cores and support a total of 452 WoodWave panels.

Throughout the design phase, two roof options were developed: the innovative but untested WoodWave system, and a more conventional steel-deck-on-glulam purlin system. Both options used the primary wood-steel arches to span the almost 330 ft (100 m) width of the main hall. While both the client and the design team favored wood, a number of technical and economic questions had to be answered prior to a final decision being made.

The design team conducted research and extensive testing to verify the performance of the WoodWave system in the areas of structural capacity, acoustics, architecture, sustainability, constructability, lighting, fire safety, maintenance and durability.

These studies concluded that the WoodWave system met all the physical criteria and provided superior acoustic performance to the more common perforated metal deck alternative. There was also the additional benefit of greater aesthetic appeal—by virtue of the warm appearance of the filigreed wood ceiling and a roof panel design that conceals the sprinkler system from view.

The primary structure of the Oval was required to be of noncombustible construction, but the roof assembly was permitted to be constructed using heavy timber elements. The WoodWave option, which used dimension lumber elements smaller than traditional heavy timber, was demonstrated by fire simulation modeling to meet the life safety and structural fire integrity provisions of the building code. This was due to the overlapping configuration of the WoodWave structure, and the large volume and openness of the main space.

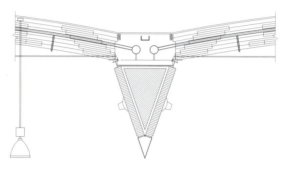

节点详图 Detail drawing

"对我而言,屋顶是最为壮观的部分,从建筑内部看来好比仰望星空。"Kristina Groves 2008 年世界杯速滑冠军说。

该栋建筑的两个较低楼层是混凝土现场浇筑而成的。其延展出巨大的倾斜的混凝土支墩,托起拱形屋顶。主要拱形梁(相距 14.3 米)是由钢桁架构件与双层胶合木构件之间,按一定的角度互相连接组合而成的。该栋建筑的机械和电气设施隐藏于这些拱形体中间。这些拱形构架之上支撑了 452 块"木浪"屋面板。

在整个设计阶段,曾有两种屋顶设计方案:一个是创新型的但未经测试的"木浪"系统,另一个是常规的胶合檩条加钢板系统。两种方案都采用了钢木混合的曲梁来横跨宽度约为 100 米宽的大厅。尽管客户和设计团队都更加青睐木材,但是在作出最后决定前仍有许多技术和经济问题有待于解决。

设计团队进行了相关的研究和广泛的试验来确认"木浪"系统在结构强度、吸声效果、建筑风格、可持续性、可施工性、照明效果、消防安全、维修性和耐久性等方面的性能情况。

经过这些研究后发现,"木浪"系统符合所有的物理标准,并具有比普通穿孔金属板更优越的吸声性能。另外,美观,也是"木浪"的一大优势——精巧的木质天花板展现出温暖优美的外观,同时巧妙地隐藏了屋顶喷淋系统。

椭圆速滑馆的主体结构必须要达到建筑防火要求,屋顶的装配只允许使用重型木结构构件。"木浪"系统使用的规格材构件要小于传统的重型木结构构件,在经过火灾模拟试验后,结果证明其符合建筑设计防火规范要求。这主要是由于具有重叠构造的"木浪"系统和具有大体积和大开敞空间的主体部分。

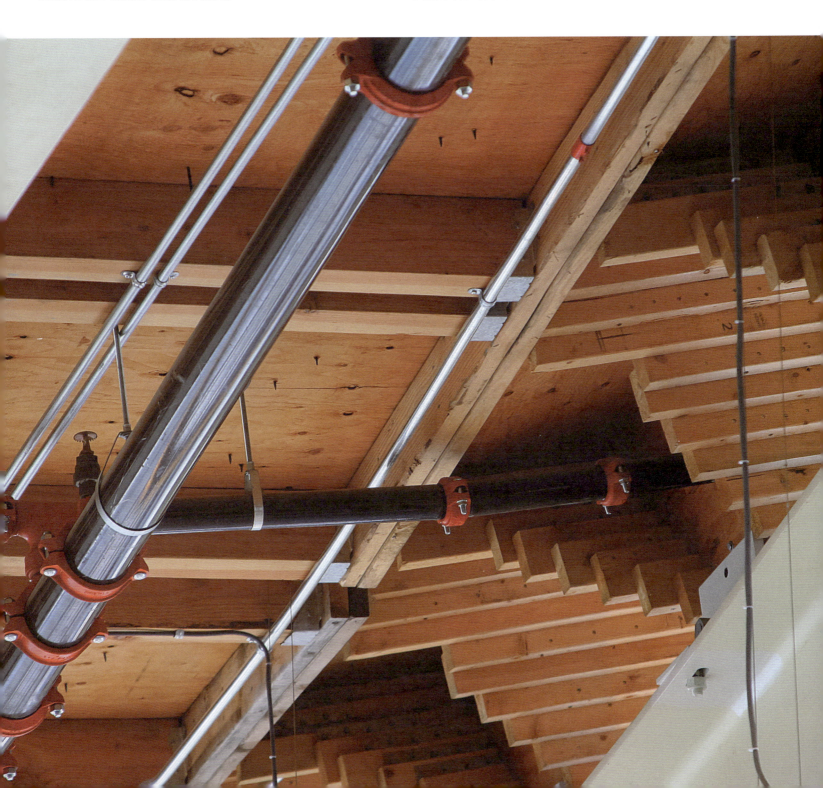

The roof uses standard materials supplied directly from British Columbia mills, including:

> 1 million bd ft (2400 m³) of 2 × 4-in (38 mm × 89 mm) spruce-pine-fir commodity dimension lumber—primarily lumber affected by the mountain pine beetle infestation in British Columbia's Interior;

> 19000 sheets of 4 × 8-ft (1.2 m × 2.4 m) Douglas-fir plywood in the roof panels;

> 1 million bd ft (2400 m³) of Douglas-fir lam stock lumber in the glulam beams

Where the roof extends beyond the walls on the north and south sides of the building, a total of 34 yellow-cedar glulam posts (29000 bd ft / 70 m³) support the overhangs. The venue features an energy-saving refrigeration plant and a state-of-the-art rain water collection system. The Richmond Olympic Oval project, which also includes a waterfront plaza, park and parkade, was completed under the budget of CAD $178 million.

"The Oval roof demonstrates that Mountain Pine Beetle wood is a good material and can be attractive. While it won't address the epidemic directly, showcasing the use of the wood will hopefullyhelp encourage its use elsewhere."
— *Greg Scott richmond's director of major projects*

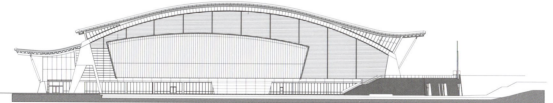

立面图 Elevation

屋顶所使用的标准材料是由不列颠哥伦比亚省工厂直接提供的，包括：

> 2400 立方米的 2×4 英寸（38 毫米 ×89 毫米）SPF（云杉－冷杉－松）规格材——大部分受到森林甲虫的侵害，但不影响其结构强度

> 屋顶板使用了 19000 张 4×8 英尺（1.2 米 ×2.4 米）花旗松胶合板

> 胶合梁由 2400 立方米的花旗松原木材组成

在该栋建筑的北边和南边，屋顶悬挑出外墙，采用了 34 根黄柏胶合木柱（70 立方米）来支撑这些突出部分。该馆的又一特色之处是其节能冷藏设备和先进的雨水收集系统。列治文奥林匹克椭圆速滑馆项目，同时还包括了海滨广场、公园和停车场，建设费用低于 1.78 亿加元的项目预算。

"椭圆速滑馆的屋顶证明了森林甲虫木材是一种很好很可靠的建筑材料并且是具有吸引力的。虽然这不能直接解决森林甲虫蔓延的状况，但是展示该木材的使用情况，或许能让更多人也来使用它。"

—— Greg Scott 列治文重大项目总监

Wood and Sustainability

The oval is designed to meet leading-edge, high-performance building standards. The structure itself is designed to qualify for Silver Certification on the Leadership in Energy and Environmental Design scale (LEED®), as well as for Green Globes.

Solid sawn lumber is by far the least energy intensive and least polluting form of construction, but prior to the creation of the WoodWave system, there was a very limited understanding of the possibilities of light framing for large and long-span buildings.

The specification of mountain pine beetle lumber from British Columbia forests and mills means the Oval's roof sets an important precedent in the effort to mitigate the impact of this unfortunate effect of climate change.

If incorporated into buildings, the salvaged wood will continue to store carbon and delay the release of carbon dioxide (a key component of greenhouse gas emissions) into the atmosphere for the life of the structure. If left to decay in the forest, this release of carbon dioxide would happen much more quickly.

Because of its low embodied energy, low toxicity and the carbon sequestered within it, wood makes a significant contribution to the overall environmental performance of the building. For example, the amount of carbon sequestered and stored in the wood used in the building amounts to 2900 metric tons of CO_2. Add to this an estimated 5900 metric tons of CO_2 accounting for amount of avoided emissions resulting from the use of wood instead of amore greenhouse gas intensive materials and the total potential carbon benefit is 8800 metric tons of CO_2. This amounts to removing over 1600 cars from the road annually or to the energy used to power 800 homes in a year.

Wood products require less energy to extract and process than other materials, and buildings that use wood can require less energy to construct and operate over time. If less fossil fuel energy is consumed, fewer greenhouse gases are emitted.

In addition to using lumber obtained from British Columbia forests, wood ceilings and paneling were milled from trees felled on the site. Cuttings were taken from trees planted when the area was owned by Richmond pioneer, Samuel Brig house. After being propagated in City of Richmond nurseries, they will be planted along the site's picturesque new Samuel Brig house Heritage Boulevard.

Designed to convert between a speed skating facility, elite athlete training zone, international sports venue, neighborhood recreation Centre and seniors' rehabilitation area, the Richmond Olympic Oval will be a center for sports of all kinds.

木材及其可持续性利用

该椭圆速滑馆的设计符合先进的高性能建筑标准。建筑本身的设计获得LEED认证银奖,符合"绿色地球"标准要求。

实木锯材是目前能耗最少并且污染最少的建筑材料,但是在创造出"木浪"系统之前,人们对于在大跨度的大型建筑中使用轻型木结构框架知之甚少。

这些产自不列颠哥伦比亚省受到森林甲虫影响的锯材的具体材料特性和锯木厂的实例表明,椭圆速滑馆的屋顶,为积极地减轻气候变化带来的负面影响,开启了一个重要的先例。

如果将其运用到建筑上,在其生命周期内,可回收利用的木材可以持续储存碳,并可延迟将二氧化碳(温室气体的主要组成之一)释放到大气中。但是如果把木材扔弃在森林里任其腐烂,那么二氧化碳会很快被释放出来。

由于木材具有低内藏能、低毒性的特征以及碳存储功能,在速滑馆中使用的木材可大大减少建筑对环境产生的负面影响。例如,用于速滑馆的木材共存储了约2900吨二氧化碳。再加上,据预计正是由于使用了木材而非其他温室气体密集型材料,使建筑减少了约5900吨二氧化碳的排放。因此,潜在的负碳排放总量为8800吨二氧化碳。这一数据相当于1600辆汽车一年的排放量,或800户家庭一年的能耗。

木材在取材和加工过程中的耗能要少于其他材料,故使用木材在建筑的建造和使用运营期间的耗能也更少。如果能减少化石能源的使用,就可以大幅减少温室气体的排放。

除了采用不列颠哥伦比亚省森林采伐的木材以外,木制天花板和镶板更是就地取材,使用了施工场地的树木。这些树木是由早期这块土地的拥有者、列治文市的先驱,塞缪尔·布里格豪斯(Samuel Brighouse)种植的。新的树苗经过里士满当地苗圃培育后,被沿着风景如画的新塞缪尔·布里格豪斯大道栽种。

速滑馆被设计成一个多功能运动中心,不仅可以提供速滑设施,还能作为精英运动员训练场地、国际体育赛事举办场地、社区康乐中心和老年人康复场馆。

"We've spanned this enormous distance with panels using ordinary two-by-fours, the same kind you can find in every house in Canada. We couldn't have designed or built this even 10 years ago."
— Gerry Epp fast + epp structural engineers

The WoodWave Structural Panel© System

The WoodWave panels are designed to span between the primary wood-steel arches. Each WoodWave roof panel consists of three parallel Vee-trusses: hollow, arched triangular sections, typically 42 ft (12.8 m) long, 4 ft (1.2 m) wide and 26 in (660 mm) deep, laid side by side and connected together by a 1 b/i-in (28-mm) thick stressed skin of plywood to form a 12 × 42-ft (3.6 m × 12.8 m) long panel.

The two sloping faces of each Vee-truss are built up from successive strands of 2 × 4-in (38 mm × 89 mm) lumber on edge, and splay out from a central bottom chord (the keel) of the same material. Each strand is vertically offset from the one below it and stitched together with nails and metal reinforcing strips. Only every second strand is continuous, with alternating strands comprised of short lengths of lumber. These splice blocks are separated longitudinally by gaps of varying length, thus creating voids that lighten the structure and enhance acoustic performance.

The incremental longitudinal offset of each successive row of splice blocks also creates trussing action along the length of the arched Vee-truss, thereby improving resistance to bending. At intervals along the span, inside the Vee-trusses, triangular plywood gusset plates improve lateral stiffness and help to maintain the precise geometry of the unit.

The shop fabrication process utilizes both custom computer numerically controlled (CNC) machinery and manual means to grade and label the lumber, produce the varying lengths of lumber strands, assemble the strands into groups, press each Vee-truss into an arch, and install the steel rod tension tie, giving it a camber of 26 in (660 mm). The final product is a composite panel whose structural performance is complex, but where each component performs at optimal efficiency.

The resultant undulating WoodWave panel soffit, the regular openings in the ceiling and the saw tooth profile of the Vee-trusses also help minimize reverberation of crowd noise and amplified sound, while the shadow pattern created by the filigreed lumber strands presents a visual texture that is rare in a building of this scale.

"如今,我们使用了最普通的 2×4 规格材来跨越如此庞大的跨度,就像是加拿大所有房子里都看得到的常规材料。在 10 年前我们都还无法做到这样的设计或建造。"
—— Gerry Epp Fast + Epp 结构工程咨询公司

"木浪"结构板系统

"木浪"设计横跨于主体钢木结构曲梁之间。每一块"木浪"屋顶板由 3 个平行的 V 型桁架组成:它们是空心的、拱形三角形构件,长 12.8 米,宽 1.2 米、深 0.66 米,三个 V 型桁架并排放置,由一块 28 毫米厚的、结构受力的胶合板进行连接,从而形成了一组"木浪"(3.6 米 ×12.8 米)。

V 型桁架的两个坡面是用竖立的 2×4(38 毫米 ×89 毫米)规格材连续成排地铺设而成,并由相同规格材制成的、位于 V 型桁架中心底部的弦杆(龙骨)处伸展出来。每一块规格材与下一块规格材之间呈垂直偏离,并且通过钉子和金属加固条"缝合"在一起。每两块规格材之间具有连续性,其中间的规格材由短木组成。这些拼接木块之间纵向间距长度不一,从而形成空隙以减轻整体结构,并提高吸声性能。

每一排纵向偏移的连续拼接木块,沿着 V 型拱形桁架的长度方向产生了约束力,从而提高了抗弯强度。沿着跨度每隔一定距离,在 V 型桁架的内部,三角形的胶合木结点加固板可以提高横向刚度并且有助于维持该构件精确的几何形状。

工厂制造程序采用了自定义的计算机数字控制(CNC)机械操作和人工手动的方式,为锯材定级标注木材,生产不同长度的木块构件,将构件分类组装、把每一个 V 型桁架压入拱梁中,安装钢材拉杆,并使其具有一个 660 毫米的拱曲度。最终的成品是一种复合板,这种复合板的结构比较复杂,各个构件都能起到最优效果。

最终制成的波浪形"木浪"板底面,在天花板处规则的开口以及 V 型桁架的锯齿状轮廓有助于最大限度地降低人群噪声和扬声器混合的声音。由细丝木条所形成的阴影效果给人一种视觉质感,这个手法在这样规模的建筑物中十分罕见。

The WoodWave roof panel system was conceived and engineered by StructureCraft Builders, a British Columbia-based design-build company. It is so named because it is built completely of lumber and plywoodfastened together in a wave-like form to serve as a structural roof, asa finished ceiling and as a noise-reducing panel.

The panels are of tied arch construction that develops composite action between the laminated lumber arch and the plywood top skin and bulkheads. The construction includes a tension tie below, which creates acamber that in turn increases the effective depth of the system.

The resulting self-contained, arched panels are simple to erect and providea stable platform for rapid installation of successive panels. Cnc prefabrication of the lumber strands and shop-based pre-assemblyensure precision and consistent quality of the final product. To meet fire codes, factory fabrication of the panels also included the installation of sprinkler lines, a black fabric acoustic liner and mineral wool insulation within the Vee-trusses.

Acoustic performance of the building is enhanced by the same mineral wool, as well as by the large area of openings in the underside of the panels. The Vee-shapes of the panel structure create internal voids for concealingthe sprinkler piping, while allowing the sprinkler heads to protrude into the space below.

The 452 WoodWave panels were pre-fabricated by StructureCraft Builders'25-member carpenter crew in 8 months and erected by StructureCraft Builders' 10-member carpenter crew.

"The Oval will provide great benefits for our community for generations to come. With its twin focus on sports and wellness, it will have programs and services that will appeal to members of the entire community, regardless of their age, fitness or physical ability."
— *Malcolm Brodie mayor of Richmond*

"木浪"屋顶木板系统是由一家位于不列颠哥伦比亚省的建筑施工单位——StructureCraft Builders 构思和设计的。"木浪"这一命名是由于该屋顶系统全然通过规格材与胶合木的组合链接，营造出波浪的形状。它既是屋顶结构，亦是精致美观的屋面吊顶，还可作为消声面板使用。

整个屋顶面板呈拉杆结构，由胶合拱梁、胶合木顶壳板及隔板互相组合而成。其结构包括下方的拉杆，它使其产生了弧度，并反过来增加了屋顶系统的有效深度。

由此产生的独立的拱形板，施工较为简单，它为快速安装连续板材系统提供了稳定的技术平台。经计算机数控（CNC）预加工的木材构件和以车间加工为基础的预拼装，确保了最终成品的精确性和稳定的质量。为了符合消防规范，工厂里制造这些木板时也包含了喷淋管线和一条黑色声音传输线的安装，以及在 V 型桁架内填充矿物棉保温材料。

相同的矿物棉还起到很好的隔声效果，"木浪"板下方的大面积开口也提高了建筑物的吸声性能。V 型结构所形成的中空空间，隐藏了喷淋管道，同时使喷淋头露出于吊顶板之下。

452 块"木浪"板由 StructureCraft Builders 建筑施工公司的 25 位木匠团队历时 8 个月内制作完成，然后由该公司的 10 人木匠团队进行安装。

"该椭圆速滑馆将给生活在我们社区的世世代代带来巨大效益。速滑馆将聚焦体育运动与健康生活，举办各种活动并提供各类服务来吸引整个社区的各类人群，不论他们的年龄、体态或体能。"

—— Malcolm Brodie 列治文市市长

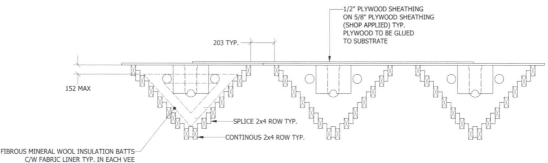

节点详图 Detail drawing

加拿大观景山庄水上中心
Grandview Heights Aquatic Centre, Canada

来源：Naturally Wood
翻译：高晓萌
校对：沈姗姗
摄影：Joshua Peter Esterhuizen, HCMA Architecture + Design

Location: Surrey, British Columbia
Size: 8830 m²
Completion: 2016
Architect: HCMA Architecture + Design
Structural engineer: Fast + Epp
Construction manager: Ellis Don
Engineered wood fabricator: Western Archrib
Glulam installation/connectors: Seagate Consulting Ltd.
Project owner: City of Surrey

Awards:
2016 Sports Award, World Architecture Festival
2016 Citation Award, Wood Design & Building Awards
2016 Supreme Award, The Institution of Structural Engineers, UK
2017 Wood Design Award – Engineer, Wood WORKS! British Columbia

"The City of Surrey is committed to building vibrant, healthy, sustainable communities and as part of that goal we have a policy to consider the use of wood in our capital projects. Because wood is a sustainable local resource and provides a sense of warmth, it fits well with these City goals."
—*Scott Groves Manager, Civic Facilities Division, City of Surrey*

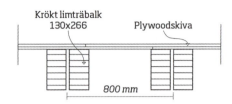

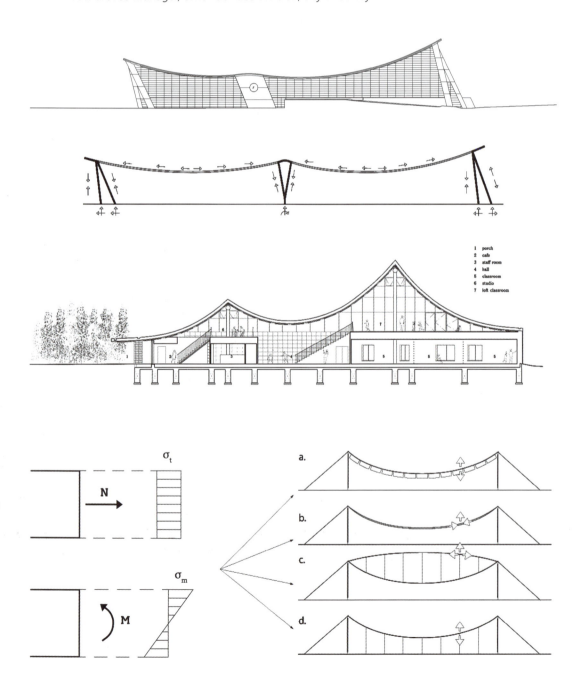

项目地点：素里市，不列颠哥伦比亚省
建筑面积：8830 平方米
竣工时间：2016 年
设计师：HCMA Architecture + Design
结构工程师：Fast + Epp
施工经理：Ellis Don
工程木制造商：Western Archrib
胶合木安装／连接件：Seagate Consulting Ltd.
项目业主：素里市

获奖情况：

2016 年　运动场场馆奖，世界建筑节
2016 年　引用奖，木设计及建筑奖
2016 年　最高奖，英国结构工程师协会
2017 年　木材设计工程奖，Wood WORKS! 不列颠哥伦比亚省

"素里市致力于建筑活力、健康和可持续的社区，作为这一目标的一部分，我们制定了一项在我们的资本项目中使用木材的政策。因为木材是一种可持续的当地资源，也给人一种温暖的感觉，这很契合我们城市的目标。"

——Scott Groves 素里市城市设施部经理

Project Overview

Located at the center of the fastest growing community within B.C.'s fastest growing municipality, this new facility is designed to accommodate both international swim meets and a wide variety of water-related community activities.

The aquatic center, with its dramatic suspended roof form, is the first project to be completed on the 'super block' that is destined to become a regional campus of health, wellness and sports excellence. In addition to its natatorium which houses a competition-sized lap pool and a leisure pool - both overlooked from the main lobby - the complex includes two hot pools, a sauna, fitness center and poolside cafe.

At one end is a diving tower, with boards and platforms from one meter to 10 meters in height. At the opposite end is a spiraling water slide positioned adjacent to the glazed curtain wall, so as to be clearly visible from the street.

These two structures create natural high points for the tensile roof, which swoops down from concrete buttresses at either end, to a transverse concrete frame that bridges the natatorium between the leisure and competition pools.

The roof form is both practical and economical. Compared to a flat roof, it reduces both the exterior surface area and the internal volume of the building, resulting in a reduction in capital cost for building materials and labor, as well as decreased operating costs related to heating and cooling. This effect is further enhanced by the shallow depth of the tensile structure compared to a conventional beam or truss system.

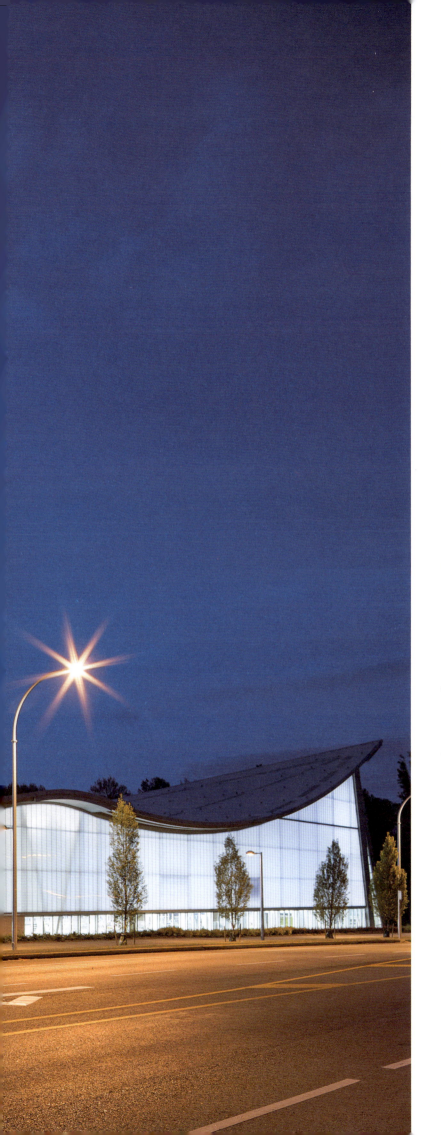

项目概况

加拿大观景山庄水上中心位于不列颠哥伦比亚省内发展最快的市和成长最快的社区，该新设施的设计是为了容纳国际游泳比赛和各种各样的水上社区活动。

水上活动中心有着引人注目的悬吊屋顶形式，是第一个在"超级社区"完成的项目，这个"超级社区"注定会成为一个健康、具有运动活力的社区。除此之外，还有一个比赛规模的游泳池和一个休闲游泳池（都可以从大厅俯瞰），综合设施包括两个温泉池、一个桑拿浴室、健身中心和池畔咖啡馆。

该中心的一端是一个跳水台，有从1米到10米高的板和平台。在另一端，是一个螺旋滑水道，位于玻璃幕墙附近，可以从街上清晰可见。

悬吊屋顶结构由两端的混凝土墩柱支承，横向混凝土框架连接了休闲游泳池和竞赛泳池。

屋顶形式兼具实用性和经济性。与平屋顶相比，它减少了建筑的外表面积和内部体积，从而降低了建筑材料和人工的成本，也降低了制冷保温有关的运营成本。与传统的梁板或桁架结构系统相比，悬吊屋顶由于构件全截面受拉，可以使得屋顶结构更轻薄。

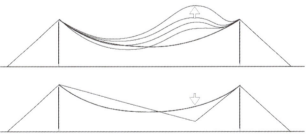

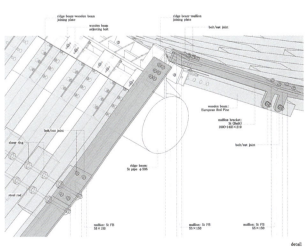

detail

Wood Use

Wood was chosen for the roof structure as it met several important design considerations, both architectural and structural. Wood structures have a proven track record in high humidity environments, as glue laminated (glulam) beam roof systems have often been used in aquatic facilities for their resistance to warping due to moisture. The natural appearance of wood also lends a warm atmosphere to facilities that, of necessity, have large areas of hard, impervious surfaces.

For the suspended roof at Grandview Heights, glulam beams also offered the required tensile capacity, self weight and inherent stiffness required to resists wind uplift, when compared to the commonly used steel cable system. Because such systems generally come with expensive proprietary connectors, the glulam system also proved to be the more economical option.

The paired 130mm x 220mm glulam beams are spaced at 700mm centers and oriented along the length of the pools below. The larger (65m) span is divided into three sections, and the shorter (45m) span into two sections. In each case the paired beams were fitted in the factory with steel knife plates at either end, enabling the field connections to be steel to steel. This is an important aspect of 'constructability', and shows how a 65 meter long catenary curve can be realized in wood. The end beams are anchored to steel plates embedded in the concrete buttresses at either end of the natatorium, and in the concrete frame that separates the spans. The roof diaphragm consists of two layers of plywood installed on site.

Suspended roofs can be subject to considerable deflection as they lack the rigidity of truss and beam structures. The design team worked carefully to ensure that the deflection of the roof under different loading conditions (most critically a heavy snow load) would not exceed 200mm – the maximum that could be accommodated at the roof edges by a standard curtain wall slip joint detail.

木材使用

屋顶结构选择了木材,因为它满足几个重要的设计要求、建筑和结构方面的考虑。木结构在高湿度环境中有良好的表现,胶合木梁屋顶结构经常被用于水上设施,因为它们能抵抗水分引起的翘曲的抵抗力。木材的自然外观也为设施提供了温暖的氛围,也有大面积坚硬、不透水的表面。

与常用的钢索结构系统相比,观景山庄水上中心采用的悬吊屋顶结构,其胶合木梁提供了所需的抗拉承载力、自重和固有刚性,来抵抗屋顶产生的风吸力。因为钢索结构通常使用昂贵的专用连接件,而胶合木结构系统是更经济的选择。

成对的 130 毫米 × 220 毫米胶合木梁间距的中对中距离为 700 毫米,沿下面水池的长度方向排列。大跨度(65 米)分为三个部分,短跨度(45 米)分为两个部分。每对梁构件在工厂加工时,将钢片板安装在梁构件的两端,使之在现场能与钢连接件连接。这是"可施工性"的重要方面,展现了弧长 65 米的悬浮屋顶如何由木材来建造。胶合木梁的两端与在游泳馆两端的混凝土墩柱的颈埋钢板连接固定。屋盖系统由两层胶合板在现场安装而成。

悬吊屋顶结构由于缺乏桁架和梁结构的刚度支撑,会有相当大的挠度。设计团队经过复核,以确保不同荷载条件下(最关键的是雪荷载)屋面的挠度不超过 200 毫米,这是标准幕墙和屋顶变形缝可接受的最大挠度。

Load-bearing structures for large spans should ideally be designed so that they primarily work in tension or compression. This allows for optimum use of the materials that make up the structure, which in turn often results in a slender and elegant form.

A strong, light sheet - It is hardly surprising that a structural component under an axial load is more effective than the equivalent component under bending stress. This can easily be appreciated by studying the distribution of stresses in a cross-section that is subject to axial force and moment, respectively. The stress under an axial load is evenly distributed, which enables full utilization of the cross-section's capacity. In the cross-section under bending stress, however, the stress shows a linear distribution. This means that the material's full capacity can only be achieved in the outer fibers of the cross-section, while the central part of the cross-section remains underutilized.

In regular suspension structures, the primary load-bearing duties are discharged by wires or cables. These lack any capacity to take up moment and therefore transfer loads only through tensile stresses in the primary load-bearing structure. In addition, wires and cables lack bending stiffness, which means that the only way these structures can bear a load is by changing their original form. In many cases, this form change can be problematic, particularly for roof structures that are subject to non-symmetrical loads, such as uneven snow load. In such a situation, a flexible suspension structure would be highly likely to suffer unacceptably large deformations. Sensitivity to wind-induced instability – i.e. wind oscillations created by gusts of wind or periodic vortex shedding – is also a common problem in flexible suspension structures. There are, however, a number of methods for stiffening up suspension roof structures and so reducing the risk of large deformations and wind-induced instability.

大跨度的承重结构，在其抗拉和抗压方面，应进行合理的设计。这样能使建造结构件的材料得到最优化使用，而这往往也能使建筑取得一个细长而优雅的形态。

对于一个强度高而轻盈的板式构件来说，一个结构组件可承受的轴向载荷比其等效组件在弯曲应力作用下更为有效，这个结果是不足为奇的。我们可以清楚地了解到，在分别受到轴向力和弯矩作用下，横截面上的应力分布情况。这个轴向荷载作用下的压力分布均匀，其充分利用了横截面的承载力。而在受弯作用的截面上，应力呈线性分布。这意味着对材料的全部承载力而言，仅仅只有横截面的外层纤维起到承载作用，横截面的中心部分没有得到利用。

在常规的悬吊结构中，主要的承重任务由索或缆来承担。这种结构形式缺乏承受弯矩的能力，因此只能通过主承重结构的拉伸应力来传递荷载。此外，这些索和缆构件缺乏抗挠刚度，这意味着这些结构能够承受荷载的唯一方式是改变它们原来的形式。在很多情况下，这种形式的改变是有问题的，特别是屋顶结构，它受到非对称荷载的影响，例如不均匀分布的雪荷载。在这种情况下，柔性悬吊结构极有可能遭受无法承受的巨大形变。柔性悬架结构对风致失稳（即由阵风或周期性旋涡引起的风振）的敏感性也是一个常见问题。因此，有许多方法可以加固悬吊屋顶结构，从而降低大变形和风致失稳带来的危险。

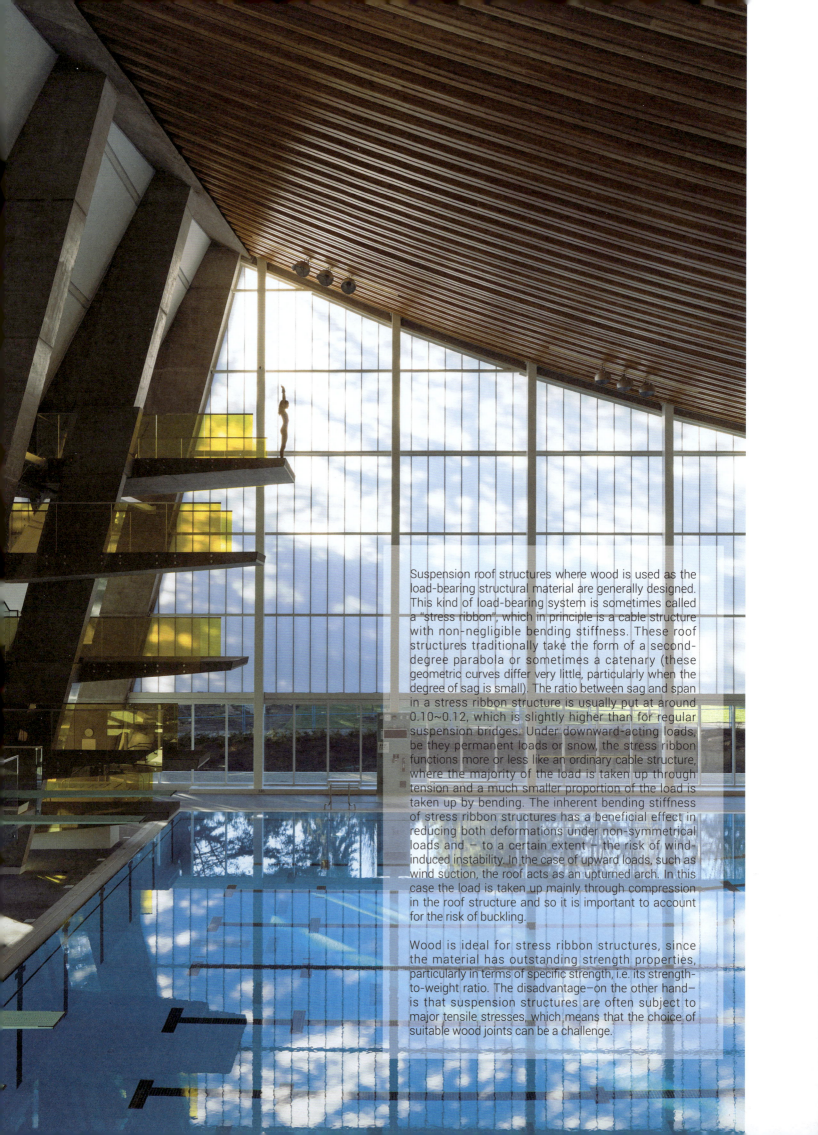

Suspension roof structures where wood is used as the load-bearing structural material are generally designed. This kind of load-bearing system is sometimes called a "stress ribbon", which in principle is a cable structure with non-negligible bending stiffness. These roof structures traditionally take the form of a second-degree parabola or sometimes a catenary (these geometric curves differ very little, particularly when the degree of sag is small). The ratio between sag and span in a stress ribbon structure is usually put at around 0.10~0.12, which is slightly higher than for regular suspension bridges. Under downward-acting loads, be they permanent loads or snow, the stress ribbon functions more or less like an ordinary cable structure, where the majority of the load is taken up through tension and a much smaller proportion of the load is taken up by bending. The inherent bending stiffness of stress ribbon structures has a beneficial effect in reducing both deformations under non-symmetrical loads and – to a certain extent – the risk of wind-induced instability. In the case of upward loads, such as wind suction, the roof acts as an upturned arch. In this case the load is taken up mainly through compression in the roof structure and so it is important to account for the risk of buckling.

Wood is ideal for stress ribbon structures, since the material has outstanding strength properties, particularly in terms of specific strength, i.e. its strength-to-weight ratio. The disadvantage—on the other hand— is that suspension structures are often subject to major tensile stresses, which means that the choice of suitable wood joints can be a challenge.

采用木材作为承重结构材料的悬吊屋面结构可按一般原则进行设计。这种承重系统有时被称为"应力带",其原则是设计一种具有抗挠刚度的索结构。这些屋顶结构,按其设计惯例,采用二次抛物线的形式,有时也采用垂曲线的形式(这些几何曲线差别很小,尤其是在凹陷程度很小的情况下)。在应力带结构中,垂度与跨径的比值一般在0.10~0.12,略高于普通悬索桥。在向下的荷载作用下,无论是永久性荷载还是雪荷载,应力带的作用或多或少类似于普通的缆结构,在这种结构中,大部分荷载是拉力,其承担的弯曲应力占很少比例。应力带结构的固有抗挠刚度对减小非对称荷载作用下的变形和风致失稳现象都有一定的作用。在向上荷载的情况下,如风的吸力,屋顶起到一个向上拱的作用。在这种情况下,荷载主要是通过顶板结构的抗压性来承担的,因此考虑屋顶结构受弯变形所带来的危险,是很重要的。

木材是应力带结构的理想材料,因为这种材料具有优异的强度特性,特别是在比强度方面,例如,强度和重量比率。另一方面,悬吊结构的缺点是主要承受较大的拉应力,这意味着选择合适的木节点可能是一个挑战。

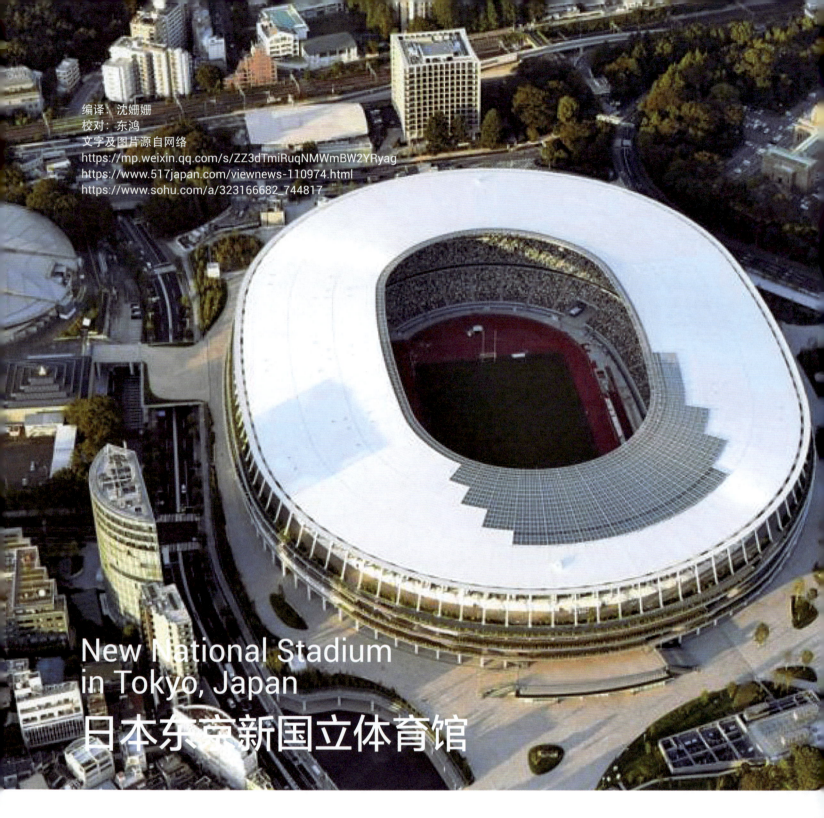

New National Stadium in Tokyo, Japan
日本东京新国立体育馆

Project address: Shinjuku-ku, Tokyo, Japan
Project function: comprehensive gym
Designer: Kengo Kuma
Cooperative Design Team: Taisei Construction

The new National Stadium was completed on November 30, 2019, and its construction had taken for about 3 years. The stadium is located in Shinjuku, Tokyo. It is the main venue for the 2020 Tokyo Olympic Games and Paralympic Games. It can accommodate 60000 spectators. There are 500 and 750 wheelchair seats being set up for the Olympic and Paralympic Games respectively. With reference to the European Football Stadium, three levels of viewing seats with different slopes were set up, and the form makes the stadium look like a mortar. At the same time, the designer also made some special designs for the upper stand to achieve the better effect of the live experience. The universal design of caring for the disabled is also a feature of the stadium. It makes full use of the space behind the mortar-shaped gradually rising, so that the elevator can reach the upper stand vertically.

项目地址：日本东京都新宿区
项目功能：综合性体育馆
设计师：隈研吾（Kengo Kuma）
合作设计团队：大成建设（Taisei Construction）

新国立体育馆于 2019 年 11 月 30 日竣工，建设工程共计历时 3 年。该体育馆位于东京都新宿区，是 2020 年东京奥运会和残奥会的主赛场，可容纳 6 万名观众（设计师为奥运会和残奥会分别设置了 500 个和 750 个轮椅观众席）。场内座席参考欧洲足球体育场，设置三层不同坡度的观众席，这种形式使得体育场的造型如同一个研钵。同时，设计师针对上层观众席也做出了一些特殊设计，让观众获得更好的临场体验。残疾人士的通用性设计也是该体育场设计着重考虑的要素之一。充分利用研钵状逐渐上升的观众席背面空间，使升降梯能垂直到达上层观众席。

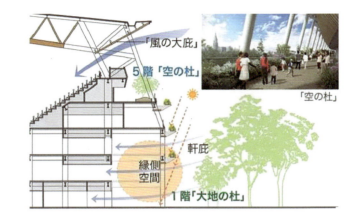

The stadium used a large amount of Japanese-made wood during its construction. The "eave of wind (natural ventilation)" structure on the roof is the design main aspect. Below the main eaves, there is also the secondary eaves around the periphery of the stadium. The wood used is produced from 47 prefectures in Japan. Through this structural design, the wind can be directed to the center of the building. A vertical wood grid is installed on the eaves, and the gap between the grids can be adjusted according to the wind direction of the season. In summer, the gap between the grid bars can be narrowed, so that the wind blowing from the south can be reached to the auditorium. Part of the wind introduced will become the updraft of the venue to take away the heat and moisture. Although there is no air conditioning in the auditorium, the natural ventilation can also cool the stadium. At the same time, considering the situation of the breeze, electric fans were installed at 185 different positions in the stadium.

The New National Stadium is the stadium that considers both the environment and versatility. It is called the "Symbiosis with the Environment". The roof structure adopts a space truss cantilever with 60m, the lower string and the pole are steel-wood structures. The main structure of the stand is reinforced concrete and steel-reinforced concrete structure. The inclined beam of the stand is built by steel-reinforced concrete, which forms a relatively stable structure.

该体育场在建设过程中使用了大量的日本国产木材。屋顶上的"风之房檐"结构是设计重点。主檐的下方还设置了一个绕体育场外围一圈的支檐，使用的木材均产自于日本47个都道府县。通过这样的构造设计可以将风引到建筑的中央。房檐上安装有竖状格条，根据当季的风向，格条之间的间隙可进行调节。夏天可以通过缩小格条之间的间隙，从而将从南边吹来的风引向观众席。引进来的一部分风会成为赛场的上升气流，从而起到带走场内热气和湿气的作用。观众席中虽然没有空调，但是利用自然风也能达到给体育场降温的效果。同时，考虑到微风天气的影响，体育场内185处地方安装有电风扇。

新国立体育馆是一个兼顾环境与通用性的新时代体育场，被称为"与环境共生的体育馆"。屋盖结构采用空间桁架悬挑约60米，下弦及腹杆为钢木结构。看台主结构为钢筋混凝土及钢骨混凝土结构，看台斜梁为钢骨混凝土，整体形成比较稳定的结构。

Project Location: Nanguan District, Changchun City, Jilin Province
Construction unit: Changchun Municipality Sports Administration
Design Unit: Jilin Architecture Design Institute Co., Ltd.
Construction unit: Daxinganling Shenzhou Beiji Wood Industry Co., Ltd.

The swimming pool of Changchun National Fitness Center is located in Nanguan District, Changchun City, Jilin Province, with a total area of 10,269 square meters. The roof of the swimming pool adopted the glulam structure. The total area of the glulam-structured roof is about 6300 square meters. The structure height of this swimming pool is about 12.7m. The main structure of the large swimming pool's roof is a single-span glulam bending stringer beam structure, with a total span of 30.5m. The small pool is a single-span wedge-shaped glulam beam structure; the entrance hall is supported by glulam bending beams.

The total length and width of the swimming pool's roof is 110.072 meters and 39.484 meters. The roof is constructed by 32 glulam bending stringer beams with the span of 30.684 meter as the main load-bearing structure. The last glulam-bending beam is fixed to the main structure of the swimming pool. At the same time, the front one is supported by 1m× 1.2m × 8.5m columns. There are 30 stringer beams in the middle. Each two stringer beams set into one group, with a spacing of one meter and a group spacing of 5.374 meters. The roofs on both sides gradually widen from back to front, making fan-shape.

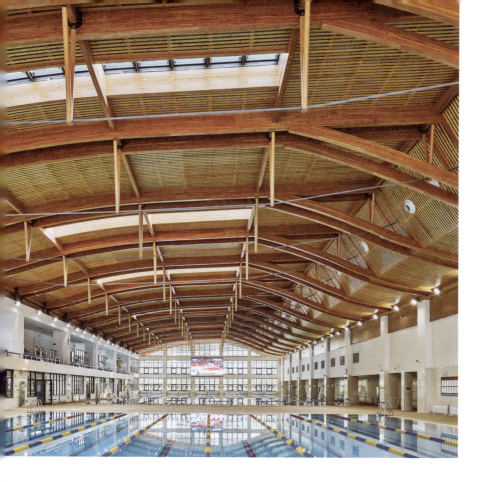

项目地点：吉林省长春市南关区
建设单位：长春市体育局
设计单位：吉林省建苑设计集团有限公司
施工单位：大兴安岭神州北极木业有限公司

　　长春市全民健身活动中心游泳馆，位于吉林省长春市南关区，总建筑面积10269平方米，游泳馆采用胶合木结构屋盖，木结构屋盖面积6300平方米，游泳馆结构高度约12.7米。其中大游泳池屋盖主结构形式为单跨张弦胶合木曲梁，沿跨度方向张悬胶合木梁总长30.5米；小游泳池为单跨楔形胶合木梁；入口门厅柱为胶合木曲梁形式。

　　游泳馆屋顶总长110.072米，宽39.484米，采用32根30.684米跨胶合木张弦曲梁做主承重结构，后端固定在游泳馆主体结构上，前端用1米×1.2米×8.5米立柱支撑。中间30根张弦梁每两根为一组，间距1米，组间距5.374米，两侧屋顶由后向前逐渐变宽，使整体呈扇面形。

长春市全民健身活动中心游泳馆

National Fitness Center in Changchun Swimming Pool, China

作者：白伟东，马建华
翻译：沈姗姗
校对：东鸿
摄影：白伟东

The bending beams with 200mm × 600mm × 8900mm set up in the front of these stringer beams, which is made of the equal number, to form the cantilever eave. It is supported by the columns with 0.5m × 1.5m × 8.5m at the back and the columns with 0.21m × 0.6m × 10.1m at the front. The gap between the stringer beam roof and the cantilever eave is 2.8m, which is used to set the roof drainage.

The entrance hall is 32.325m wide and 5.6m high supported by the five groups of tree-shaped structural components. It not only shows the beauty of the structural form, but also demonstrate the diversified applicability of the timber structure.

The two ends of the stringer beams are fixed to the concrete structure of the building's main body with metal connections. Each stringer beam is supported by an inclined beam with 130mm × 300mm × 14115mm. The inclined beam is fixed to the building body at one side, and fixed to the metal connection of the stringer beam at the other side. It makes a slot at the joint between the stringed beam and the inclined beam, builds a 10mm thick steel plate inside, with holes (diameter with 20mm, spacing with 80mm). The corresponding holes are also made in the glulam beams, and then fixed by stainless steel pins with Φ20 × 200mm.

The timber components used in the project are all connected by steel fitting. Among them, the connections of the glulam columns adopt a new patented technology—assembled tendon connection nodes. Large timber components are connected by built-in steel plates and fixed with stainless steel pins. It improves the accuracy of installation and structural strength. Beam and column joints are connected by steel plate bolts, which greatly improves site work efficiency. The small timber components are connected with the main beams and columns by bolts, wood screws, etc., which is easy to install and convenient for construction.

张弦梁前部为同等数量200毫米×600毫米×8900毫米曲梁构成的挑檐，后端使用0.5米×1.5米×8.5米立柱、前端用0.21米×0.6米×10.1米支撑。张弦梁与挑檐间距2.8米，用于设置屋顶排水设施。

门厅宽32.325米，高5.6米，使用的5组仿生树形支撑结构不仅体现出了结构形态美，更展示了木结构承重体系的多样化适用性。

张弦梁两端用金属件固定在混凝土建筑主体上，每根张弦梁下使用一根规格130毫米×300毫米×14115毫米斜梁做支撑，斜梁一段固定在建筑主体上，另一端与张弦梁用金属件连接，连接方式为在张弦梁与斜梁连接处开槽，内置10毫米厚钢板，钢板上开孔（孔径20毫米，间距80毫米），对应部位的木梁上同样开孔，再用Φ20×200毫米的不锈钢销连接固定。

项目木构件连接均采用钢连接件，其中胶合木柱脚采用新型专利技术——装配式植筋连接节点，大型木构件连接采用内置钢板，用不锈钢销固定，提高了施工安装精度和结构强度。梁柱节点通过钢板螺栓连接，现场安装效率大大提升。小型木构件与主体梁、柱采用螺栓、木螺钉等形式连接，安装简便，利于施工。

澳大利亚班吉尔广场
Bunjil Place in Casey, Australia

作者：Sara Bergqvist
校译：沈姗姗
摄影：John Gollings, Trevor Mein, Andrew Chung, FJM

Architectural firm Francis-Jones Morehen Thorp (FJMT), which has offices in Sydney and Melbourne in Australia and Oxford in the UK, focuses on innovative design for public buildings and spaces. Their work explores the interplay between building and place and links the spatial with the organic. The firm has won a long list of awards for its projects, including Bunjil Place.

Client: City of Casey
Structural engineer: Multiplex
Glulam manufacturer: Hess Timber, Germany
Size: 24500 square metres
Cost: AUD 125 million

The gigantic roof at Bunjil Place is shaped like two eagle wings, a billowing structure in glulam and steel that defines the whole building. Inspired by local Aboriginal mythology, the design connects the rapidly growing area with its history.

Casey City on the outskirts of Melbourne is Australia's fastest growing municipality. An area that was rural in nature not so long ago is now home to more than 300000 people. In order to meet the wishes and needs of current and future residents, the local authority began a project to identify what was missing from the area.

What they found was a demand for cultural facilities, meeting places and arenas, a library and functions for local services and information. On this basis, the local authority launched an architectural competition to design a concept that would meet these needs. The winner was the innovative practice FJMT, which has been responsible for several groundbreaking solutions for the public space.

"Our vision was to truly gather all the required functions under one roof. This creates a relevant contemporary solution where we bring together and unite people and build a sense of community in multicultural Casey City." says Richard Francis-Jones, Design Director at FJMT and chief designer for the project.

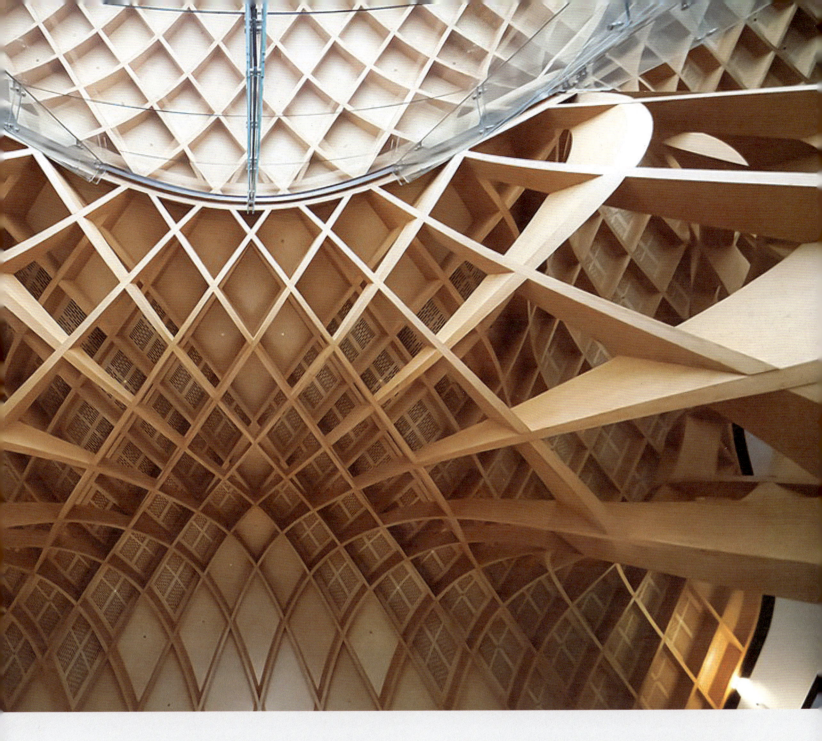

　　Francis-Jones Morehen Thorp (FJMT) 建筑事务所，其办公室位于澳大利亚的悉尼、墨尔本和英国的牛津。事务所专注于公共建筑和空间的创新设计，发现建筑与空间之间的互动关系以及空间和有机生态之间的联系。事务所的设计项目已获得多个奖项，其中包括澳大利亚班吉尔广场。

客户：凯西市
结构工程师：Multiplex. 事务所
胶合木制造：Hess Timber, 德国
建筑规模：24500 平方米
造价：1.25 亿澳币

　　班吉尔广场的巨大屋顶的形态就像两只老鹰的翅膀，向上鼓起的钢木混合结构屋顶定义了整栋建筑的形态。其灵感来源于当地的土著神话，此设计与当地快速增长区域的发展历史相关联。

　　凯西市位于墨尔本的郊区，是澳大利亚发展最快的城市。这块区域不久之前还是农村地区，现在已是超过 300000 人的家了。为了满足现在和未来居民的需求和愿望，地方当局开始了这个项目，以此来找回此区域内曾经拥有的东西（文化）。

　　当局发现本地需要文化设施、会议活动场所、图书馆，和具有本地服务和信息功能的设施。基于此，地方当局开始了一场可以满足这些需求的建筑概念设计竞赛。获胜者是具有创新实践经验的 FJMT 建筑事务所，他们对公共空间的设计，提出了多项具有突破性的解决方案。

　　"我们的设想是将所需的功能全部汇聚在一个屋顶之下。这个设想创造了一种互为关联的具有当代意义的解决方案，即在具有多元文化的凯西市中聚集和团结人民，并建立一种社区意识。" FJMT 的主创设计师 Richard Francis-Jones 说。

A collaboration with the local Aboriginal population was set up early on, which helped to give the building context and connect it to the history of the place. One of the most significant Aboriginal legends from the area is about the eagle Bunjil, the creator who protects the land and welcomes guests, but who also requires guests to follow Bunjil's laws and not harm the children or the land, which belong to him.

"The story inspired us to design a building with a roof that replicates the shape of a wing span. And in two places we've made the wings sweep down and touch the ground." explains Richard Francis-Jones.

The monumental roof comprises two parts, both of which make a strong contribution to the character of the building. The load-bearing ceiling is constructed as a grillage, a lattice of curved glulam timbers left exposed to give a warm and welcoming feel. On top of this rests the steel roof. Around the inside run mass timber beams that act as the connection between ceiling and exterior roof. There are also intermediate steel plates that help to distribute the forces. The glulam structure then continues down towards the walls, where it switches to concrete, with the load transferred via wooden beams and adjacent steel frames.

"Although the roof and ceiling are structurally connected, they also have a high degree of independence. The glulam lattice is fully self-supporting. The steel roof contains its own load-bearing elements, but is also supported by the concrete walls." says Richard Francis-Jones.

Either side of the entrance, the wooden grid of the ceiling drops all the way down to the ground at two points, like two legs that taper off and end in stylised talons. In front of the two legs stands the 12 metre-high glass facade, comprising self-supporting curved panes of glass stacked on top of each other and joined using structural adhesive. It is also here that the roof opens up towards the visitors and the building veers off inside to leave an open space outside that forms another meeting place.

Inside the glazed entrance is a large open area that can be used for many different purposes – flexibility of use is a theme that runs throughout the building. As well as the more permanent parts such as the library, café, conference room, theatre, gallery and customer service centre, there are also meeting rooms, meeting places and stages that can be temporarily used for events such as conferences, weddings, dinners, exhibitions, talks, circus and theatre.

Alongside the glulam grid, which is made of Scandinavian spruce, there are also several other visible details in wood. The wooden walls that frame both the foyer and the 800-seat theatre are clad in veneer from a single tree, a huge blackbutt, which is a kind of eucalyptus tree. The tree was found at a sawmill in Queensland that had been saving it until the right project cropped up, which it eventually did.

与当地土著居民的合作从很早就建立起来了。这有助于建立建筑语境及其与当地历史的联系。其中，一个最重要的土著传说就是关于鹰帮班吉尔的。帮主保护土地，欢迎客人，并要求客人遵守班吉尔的法律，不伤害属于他的孩子和土地。

"这个故事启发了我们去设计一座屋顶仿效翼展形态的建筑。在两个地方，我们已经设计了向下飞翔，并接触到地面的翅膀形态。" Richard Francis-Jones 解释到。

巨大的屋顶由两个部分组成，这两部分都对突出建筑特色起到了重要的作用。承重的天花板被建构成一种格栅形式，一个暴露在外的呈曲面的胶合木格栅，给人一种温暖而好客的感觉。在这个结构之上是钢结构屋顶。围绕着屋顶内部空间建构大量木梁，用以连接天花和屋面。同时，中间的钢板构件协助分散应力。胶合木结构向下延续至墙体，之后转换为混凝土结构，其荷载通过木梁和相邻的钢架结构转移至混凝土结构。

"虽然屋顶和天花在结构上是相连的，他们同时也具备高度的独立性。胶合木格栅是完全自支撑的。钢结构屋顶具有自受力构件，同时由混凝土墙体支撑。" Richard Francis-Jones 说。

入口的两边，木格栅的天花逐渐下垂到地面，汇聚在两个点的位置上，它就像慢慢变细的两条腿，最后变成爪子的形态。在两条腿的前面，矗立着 12 米高的玻璃外墙，它由自支撑的曲面玻璃板组成（玻璃板之间相互叠放，并使用了结构粘合剂连接）。也就在这里，屋顶空间向游客敞开，建筑向内偏移，对外留出一个开放空间，这个开放空间变成了另一个会议空间。

玻璃入口内侧是一个巨大的开放空间，它有多种用途——使用功能的灵活性是贯穿于整个建筑设计的一大主题。除了类似图书馆、咖啡厅、会议厅、剧院、画廊和游客服务中心这样的永久性场地以外，还有会议室、会议场地和舞台空间等用于临时活动的场地，类似举办会议、婚礼、晚宴、展览、会谈、马戏团表演和剧院等活动。

除了由斯堪的纳维亚云杉制成的胶合木网格外，木材中还有其他一些可见的细节。门厅和 800 个座位的剧院的木制墙体，都是用一棵巨大的黑基木（一种桉树）的单板覆盖的。这棵树是在昆士兰的一家锯木厂发现的。该锯木厂一直在保留着这棵树，一直等到合适它的项目出现。现在，它终于被使用了。

"The exposed wood on the walls and in the glulam structure creates the warm and welcoming feel that we wanted, in combination with a sense of quality." states Richard Francis-Jones.

The structure, with its extensive wooden grid system, is unlike anything else in Australia – and has few equivalents globally. In the search for a supplier that was able to handle the complex assignment, FJMT turned to Hess Timber in Germany, where the glulam elements were prefabricated. Every curve in the grid comprises up to 120 laminated lengths of spruce, 6 millimetres thick, with a tolerance of half a millimetre between them. A specialist tool was used to cut out the thin pieces: a six-axis robot that can cut three-dimensional shapes. The connections between the different parts of the grid were then designed to handle the structural pressure on each joint.

"Hess Timber has mainly used what they call Pitzelhängare – concealed hangers with two large locking wood screws between the primary long beams and cross-beams." says Richard.

The next challenge was about transporting the fragile elements safely.

"To make it work, detailed drawings were produced to show how everything should be packed, not unlike an IKEA flatpack, to make transport as safe and efficient as possible. After three months of transport in containers, they arrived undamaged and then we had a team of local craftsmen to put it all together."

Initially the building went by the working name of the Casey Cultural Centre, but after a while it became increasingly obvious that it should be given the same name as the mythological eagle it resembles. Now everyone in Casey City knows Bunjil Place. The building is also conveniently located near the motorway and alongside a development with a shopping centre, sports facilities and a swimming pool.

"Many of the people who go shopping on the other side now also stop off here. It's fantastic to see the way everyone who lives here has really embraced the new building as part of the community." concludes Richard Francis-Jones.

"胶合木结构墙体上的暴露式木构件创造出一种我们想要的温暖而好客的感觉,并融合了一种品质感。"Richard Francis-Jones 说。

这种木网格系统结构,与澳大利亚的其他建筑不同,在全球范围内几乎没有类似的建筑。FJMT 在寻找一家能够处理复杂任务的供应商时,选择了德国的 Hess Timber 公司,这家公司的胶合木构件是预制的。在网格中的每条曲线包含了 120 毫米高、6 毫米厚、半毫米公差的云杉。以一种专门的工具来切割木薄片:一个可以切割三维形状的六轴机器人。网格的不同部分之间的连接件被用来承受每个连接处的结构压力。

"Hess Timber 公司主要采用名叫'Pitzelhängare'的产品,即主长梁和横梁之间带有两个大锁紧木螺钉的隐藏式吊架。"Richard 说。

安全运输易碎物品是下一个挑战。

"为了使其发挥作用,制作了详细的图纸,以显示所有东西应如何包装,像宜家的扁平包装,使运输变得尽可能安全和高效。在三个月的集装箱运输后,他们安全地到达目的地,他们有一个当地的施工团队将构件拼接在一起。"

最初,建筑的工程名字为凯西市文化中心,但是过了一段时间,越来越明显的是,它应该与神话中的鹰同名。现在凯西市的每一个人都知道了班吉尔广场。这个建筑也位于高速公路附近,旁边还有一个开发区,有购物中心、体育设施和游泳池。

"很多去另一边购物的人,现在也在这里停留下来。很高兴地看到,住在这里的每个人都已经把新建筑作为社区的一部分了。"Richard Francis-Jones 总结到。

重庆龙湖两江长滩原麓社区中心
Yuanlu Community Center in Chongqing

文字：李杰
整理：谢白莎
翻译：上海成执建筑设计有限公司
校对：沈姗姗
摄影：Prism Images, Arch-Exist（三棱镜建筑空间摄影，存在建筑）

Project: Yuanlu Community Center
Architects: Challenge Design
Website: http://www.challenge-design.com
Location: Next to Longxing Ancient Town, Chongqing, China
Lead Architects: Jie Li, Wei Huang, Fang Yan, Xueyan Wu, Yangfeng Xu
Design Team: Wubing Feng, Xitao Liu, Yin Liu, Wenlong Zheng, Chengzong Xue, Yuanyuan Yan
Timber Construction Design: Dilong Chen, Juan Li, Yu Zhang
Timber Construction: JAZ BUILD
Size: 4000 m²
Completion: July, 2018
Client: Longfor Group Holdings Limited

项目名称：重庆龙湖两江长滩原麓社区中心
建筑事务所：上海成执建筑设计有限公司
事务所网站：http://www.challenge-design.com
项目详细地址：重庆市渝北区龙兴商圈
主创建筑师：李杰，黄伟，严芳，吴雪艳，许阳峰
设计团队：冯武兵，刘喜桃，刘寅，郑文龙，薛成宗，颜袁原
木结构施工图设计：陈迪龙，李娟，张宇
木结构施工：上海隽执建筑科技有限公司
建筑面积：4000 平方米
项目完成年份：2018 年 7 月
业主：重庆两江新区龙湖新御置业发展有限公司

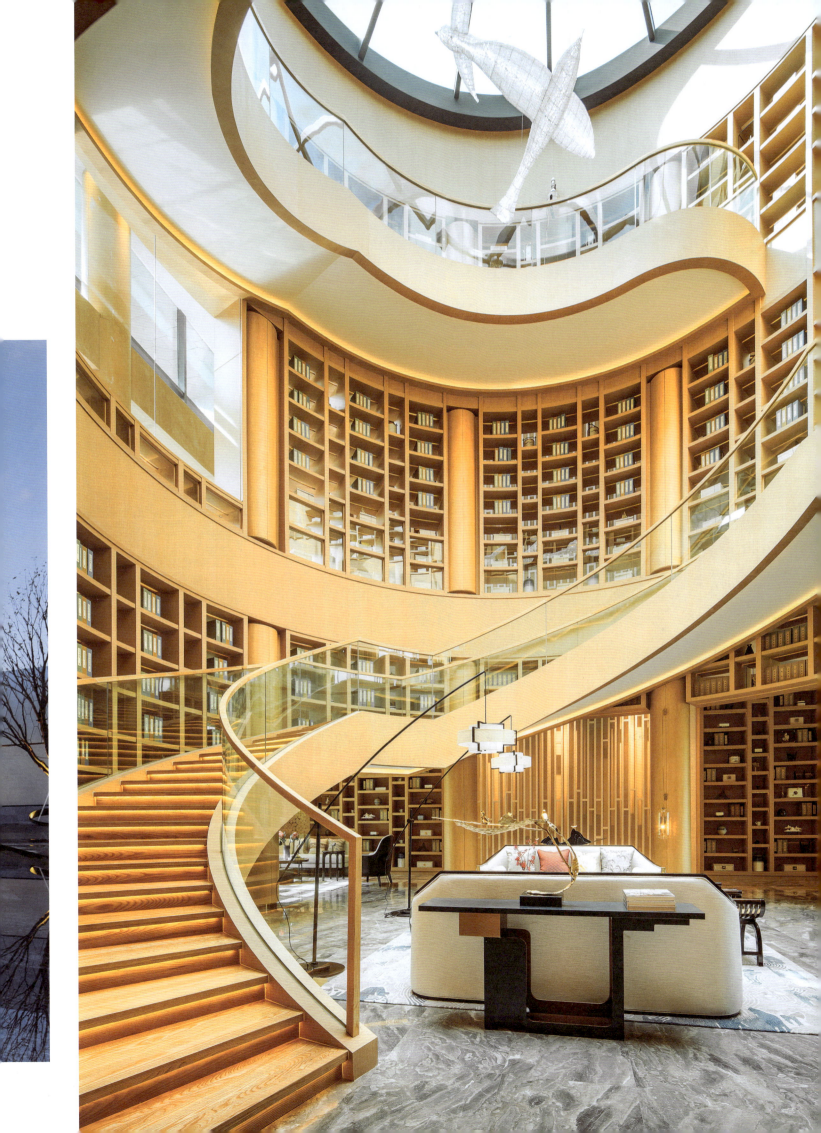

I Symbiosis

Facing Yulin River, the Project sits next to Longxing Ancient Town, Chongqing, China, mirrored by rolling hills in the distance and a centralized green area. Natural landscape on the east side is exceedingly fascinating while its counterparts on the other sides are relatively ordinary.

As a response to the environment, the architect places three buildings of different sizes side-by-side on the hillside and vertical to the riverside road, looking to achieve the optimum visual effect of riverfront landscape and distant mountains.

壹 共生

项目毗邻龙兴古镇，御临河边，有远山起伏，绿化集中，东侧景观极佳，而其余方向则乏善可陈。

作为对环境的回应，建筑师将三个体量不同的盒子并列于山坡上，垂直于滨河道路，以期达到滨河景观及远山的最佳观赏。通过建筑与环境的互塑共生，使远山、河流、庭院、建筑重塑构成新的秩序，营造独有坡地特征的建筑群落。

II Topology

Each space is designed with various sizes by the architect according to its functions, including exhibition hall, book bar, swimming pool, and restaurant, forming multiple sequential courtyard spaces. Besides, traditional Chinese artistic concepts are integrated into the modern space by virtue of the transition between virtuality and reality, which is generally applied in traditional Chinese gardens.

In addition to the interesting correlation between the spaces and sizes, changes in elevation enriches spatial forms, which enable people to experience narrative feelings in the space changed up and down, inside and out, near and far.

Courtyards and patios can evoke a sense of territory and the adjacency of inside and outside through covers, showing spatial openness, and a sense of freedom based on spatial relationships.

贰 拓扑

建筑师依据功能将每一个建筑空间依序形成不同的体量，展览厅、书吧、游泳池、餐厅等顺次形成多重院落空间，虚实空间的变化过渡犹如中式游园，将传统中式意境融入现代空间中。

空间与建筑体量形成有趣的关系，垂直变化的高度更加丰富了空间层次，上下、内外、远近，空间往复变化，人在其中，体验叙事性的情怀。

庭院与天井以覆盖获得领域感，内外场所的邻接性，并展示出空间的开放性以及基于空间关系的解放性。

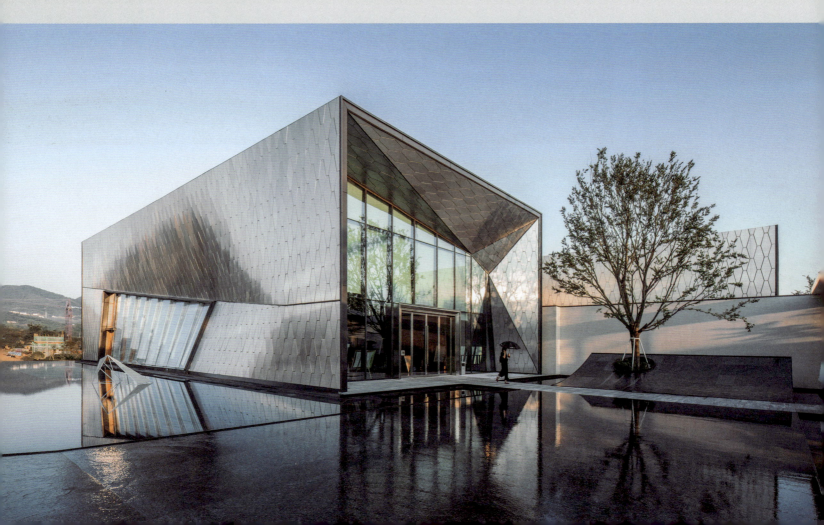

III Authenticity

In order to demonstrate the philosophy of authenticity of architecture, space, form and structure, architects use timber to closely link space and structure, thus making the architecture itself have the "virtue of honesty".

Exposed Glulam structures are applied as the crucial visual element for wooden buildings. The order and form similar to the sloping roofs in Chongqing are adopted for expression. Architects convert varying units into structural components and arrange those components in a certain changing mode, so that the architectural form changes with the space.

Wood components are lapped straight and well-structured in elaborate layout, with a strong sense of construction. Smooth lines and a sense of unity make the building space extremely charming. Dynamic balance of components presents architectural features and livens up the space.

叁 真实

为表达建筑、空间、形体的真实性哲学理念，建筑师利用木结构的特色，以结构形态塑造建筑空间和形体，将空间与结构产生紧密联系，因而使建筑本身具有了"诚实的美德"。

这种明晰的结构逻辑，试图创造一种崭新的带有乌托邦理想的建筑群落，以建筑手法对抗虚假浮华的世界。

暴露的胶合木结构构件是建筑中最重要的视觉元素。建筑师以重庆传统的折坡屋顶秩序的形式语言转换成为建筑形体，抽取变化的单元模式转为结构构件，多片构件以秩序和变化排列组合，产生紧密的空间与形体的变化。

木构件笔直搭接，严整清晰，建构感强。流畅的线条和统一的韵律感使建筑空间极具张力。构件体现出的动态平衡，丰富了建筑语言，将空间变得非常戏剧化。

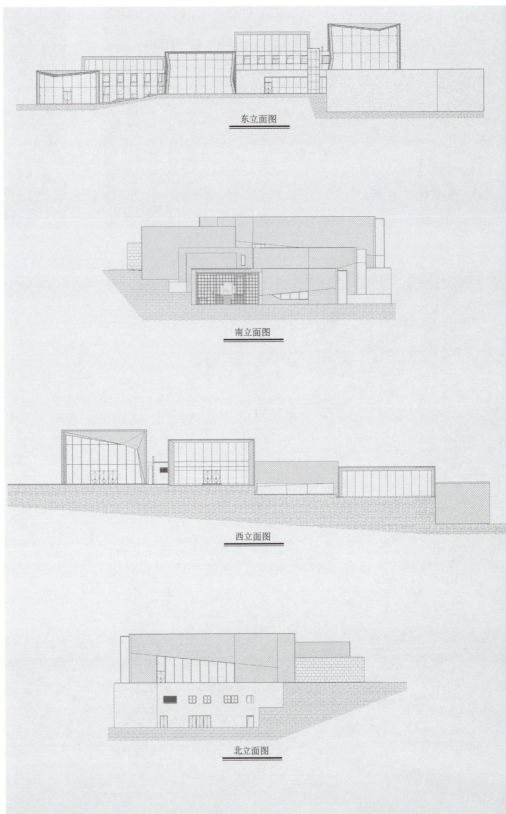

东立面图

南立面图

西立面图

北立面图

IV Circumgyration

Winding stairs and surrounding bookshelves in the book bar build a bridge that links people's minds with books. With natural light introduced from the roof, the space is full of mobility, flexibility, and the sense of wonder; furthermore, changes in light may also remind readers of meandering time.

肆 回转

设计师将书院定义成一个能让人阅读、思考、安静下来的场所，并希望成为阅读者的场所记忆，生活方式通过书院的一呼一吸中不断进行新陈代谢。书院中旋转迂回的楼梯，书架环绕，顶部引入自然光，呈现流动灵活却有震撼力的空间，光线的变化还使读者感受到时间的游走。

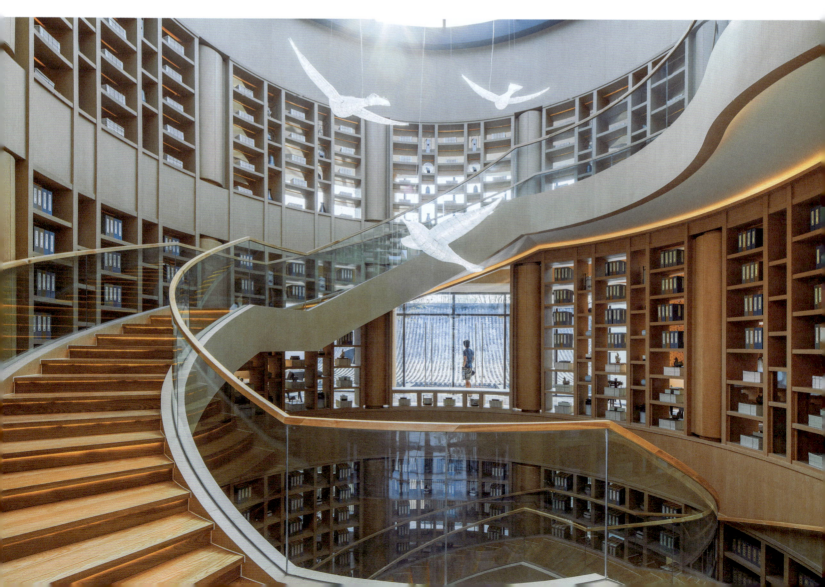

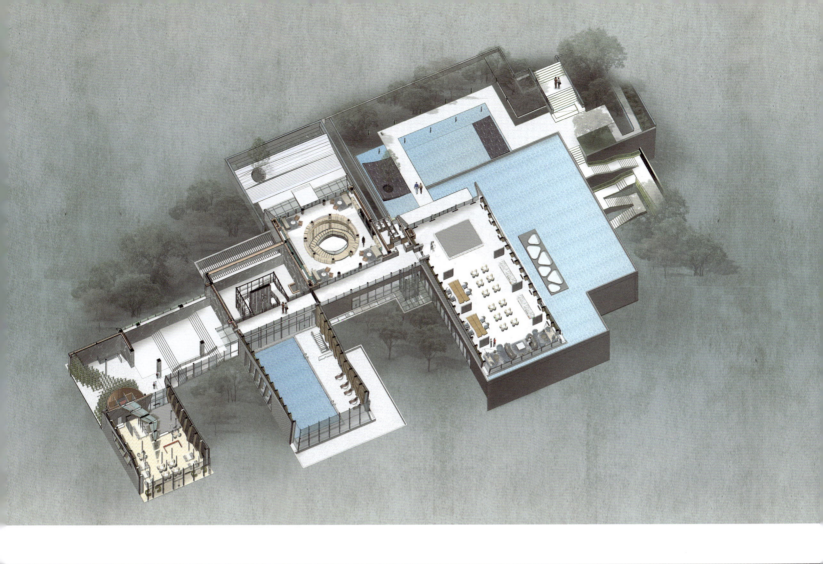
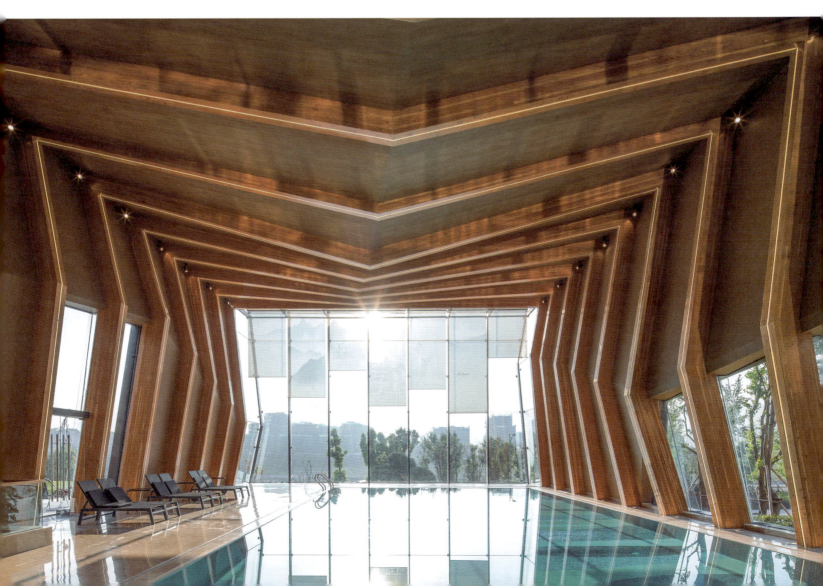

V Shadow

Adopted as the main interfacial material, hexagonal aluminum plates are covered on the interior structure by architects. Building surfaces are adaptive to changes in building form; each aluminum honeycomb plate is finely wrapped, slotted, and spliced. Digital manifestation produces a modern sense of "Cyberpunk"; the surface characteristics appear in natural light, and complicated and mysterious visual effects are generated as the light changes.

伍 光影

建筑师以六边形铝板作为主要界面材料，覆盖于内部结构之上。表皮顺应真实的建筑形态变化，每块蜂窝铝板精细包边，开缝拼挂，细节考究。数字化的表现手法营造出"Cyberpunk"（赛博朋克）的超时空现代感，表皮在自然光下形塑特质，追随光线的变化，产生复杂而神秘的视觉效果。

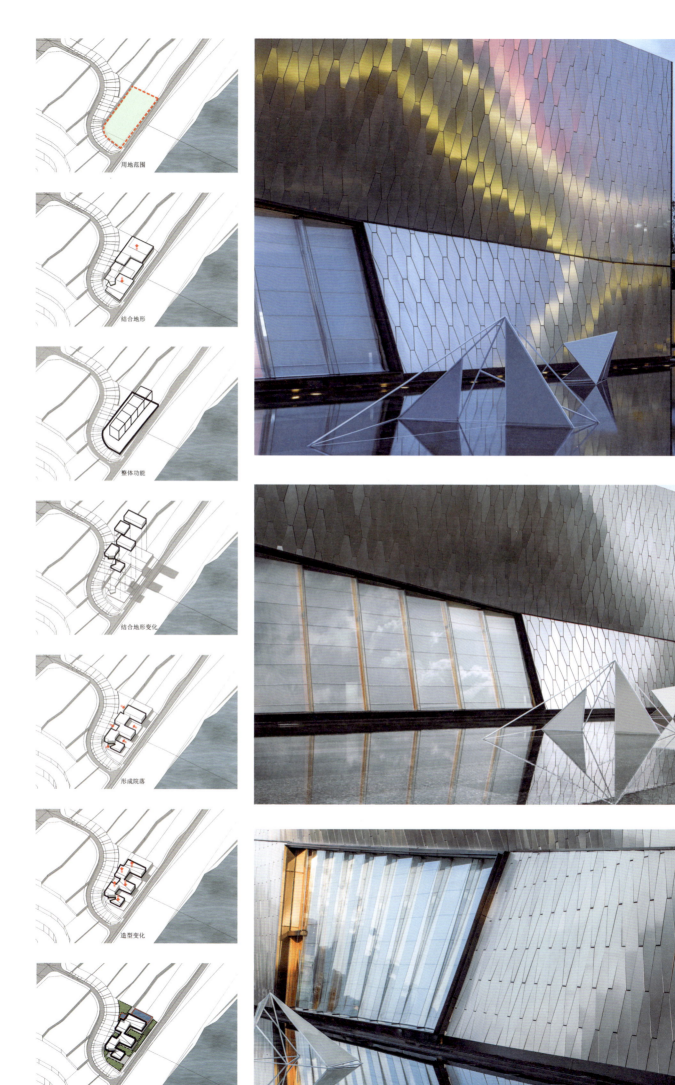

VI Technology

In pursuit of ultimate architectural beauty, design and construction are integrated for the Project. The BIM system is adopted to achieve overall project control. Each structural component is optimized thanks to programming design, manufactured by CNC machines, and installed on site. Digitization concepts and advanced technologies are applied to pre-control and dynamically manage the Project. With 3D positioning, very few errors occur during installation. It takes only 25 days from the installation of the wooden structure to the completion of the roof structure.

陆 工艺

为追求极致的建筑美观度，项目采用设计施工一体化方式，并应用 BIM 系统实现了项目整体管控，每一个结构构件都通过编程设计优化，数控机床加工生产，现场安装。为了降低施工误差，运用大量数字化理念和先进技术，将项目做到事前控制和动态管理，因此木构部分从安装到屋面结构完成仅用了短短 25 天。

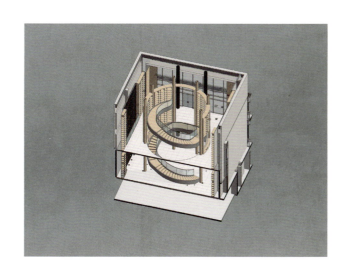

As architects have reinterpreted modern architectural aesthetics by designing authentic structure and space in an extremely modern way, the integrity and authenticity of the building from inside to outside will make people feel its beauty of morality and integrity. Furthermore, based on the changing structural form, the integration and interaction of functions, spaces, materials and structures with people are gradually unfolded, thus to exhibit the elegance of the building in natural light!

建筑师以极现代的手法，真实的结构与空间，重新阐释了现代建筑美学，这种发自于内并自内而外的建筑，以其整体性与真实性，使建筑具有了道德和诚实的美感，这种基于变化中的结构形态将功能、空间、材料、结构与人的交融和对话逐级展开，使建筑的一切都在自然的光线下优美地表现出来！

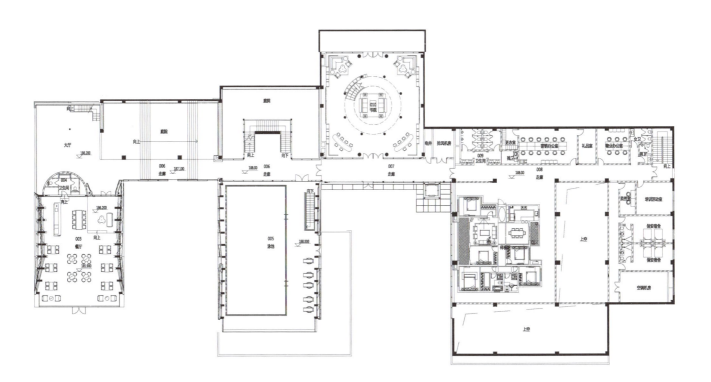

加拿大太平洋自闭症家庭中心
Pacific Autism Family Centre, Canada

来源：Naturally Wood
翻译：高晓萌
校对：沈姗姗
摄影：Derek Lepper
Credit: Derek Lepper Photography, courtesy naturallywood.com

Location: Richmond, British Columbia
Size: 5600 m²
Completion: 2016
Architect: NSDA Architects
Structural engineer: Fast + Epp
General contractor: Ventana Construction Corporation
Engineered wood fabricators: Western Archrib (glulam)
Redbuilt™ (TJI)
Project owner: Pacific Autism Family Centre Foundation

"We wanted to use wood and its inherent warmth and beauty to reinforce the welcoming atmosphere we were trying to create for people and families living with Autism. In addition, we are firm believers that wood, if properly managed, is a natural, renewable product for the long term."
— *Larry Adams, Principal, NSDA Architects*

Estimated Environmental Impact of Wood Use

Volume of wood products used: 662 cubic meters
U.S. and Canadian forests grow this much wood in: 2 minutes
Carbon stored in the wood: 601 metric tons of CO_2
Avoided greenhouse gas emissions: 1003 metric tons of CO_2
Total potential carbon benefit: 1605 metric tons of CO_2

GHG Emissions Are Equivalent to: 339 cars off the road for a year
Energy to operate 169 homes for a year

*Estimated by the Wood Carbon Calculator for Buildings, cwc.ca/carboncalculator.
*CO_2 refers to CO_2 equivalent.

项目地点：列治文，不列颠哥伦比亚省
建筑面积：5600 平方米
竣工时间：2016 年
设计师：NSDA Architects
结构工程师：Fast + Epp
总包：Ventana Construction Corporation
工程木供应商：Western Archrib（胶合木）
Redbuilt™（桁架搁栅）
项目业主：太平洋自闭症家庭中心基金会

"我们想使用木材和它与生俱来的温暖和美观，为与自闭症生活的人和家庭增添热情好客的氛围。另外，我们坚信，如果管理得当，从长远来说木材是一种天然可再生产品。"
—— Larry Adams NSDA 建筑事务所主席

使用木材预计可带来的环境影响

木材制品使用量：662 立方米
美国和加拿大森林生长出该体量的木材需要：2 分钟
木材存储碳量：601 公吨的二氧化碳
能够避免的温室效应气体排放量：1003 公吨的二氧化碳
总计潜在的碳固存量：1605 公吨的二氧化碳

减少的温室气体排放相当于：339 辆汽车行驶一年的排放量
支撑 169 个家庭运作一年的能量

* 由"建筑用木材的碳固存计算手册"估算，cwc.ca/carboncalculator.
* 二氧化碳指二氧化碳当量

Project Overview

The primary purpose of the Pacific Autism Family Centre (PAFC) is to consolidate state-of-the-art resources and research into a 'knowledge hub' to better address the growing challenge of Autistic Spectrum Disorder (ASD) in British Columbia. The PAFC will be connected to smaller satellite facilities in a network designed to build capacity for learning, assessment, treatment and support services for individuals and families across the province.

ASD is a spectrum disorder that affects the development of the brain. It is highly variable in the degree to which it impacts an individual's needs, skills and abilities. According to Sergio Cocchia, President of the Pacific Autism Centre Foundation, ASD occurs in 1 in every 69 births in British Columbia, making it the most common neurological disorder in children.

The new three-story, 5600 square meter building is located on Sea Island, close to Vancouver International Airport and flanked on either side by existing commercial buildings. The program for the facility includes a Knowledge Centre, Information Centre, Lifespan Centre and Training Centre, which together provide resource, education and recreation facilities for clients of all ages, as well as administrative and research space for the staff who support them.

In the publicly accessible parts of the building, circulation and waiting areas are deliberately oversized to prevent feelings of claustrophobia or confinement, and interiors are simply detailed to encourage a calm environment. Transparency is used strategically, with exterior views to the surrounding landscape assisting with orientation and interior views between adjacent spaces (such as corridors and stairs) to assist navigation within the building.

Wood Use

On this project, the choice of wood met all the design criteria, offering a cost effective structural solution with long spans that could accommodate future reconfiguration should the needs of autism research and treatment change.

The basic structure is a glue-laminated (glulam) post and beam frame system, supporting floors that are a combination of prefabricated nail-laminated timber (NLT) Panels and engineered light truss joisted floors, both finished with plywood decking.

The glulam posts are laid out on a 6mx6m grid for maximum economy and to ensure flexibility for future reconfiguration of the non-loadbearing partitions. On the ground floor, the posts are 2100 mm^2, and while those on the upper floors are smaller as they carry less weight. To minimize the effect of cross grain shrinkage over the height of the building, columns are superimposed one on top of the other, separated only by a steel spacer the same thickness as the concrete floor topping. The spacer forms part of a connection detail that includes saddles that carry the floor beams on either side.

The NLT panels are used in the public circulation areas and their soffits exposed for maximum visual impact. The three elevator shafts are also constructed using NLT, while the stairs are constructed with plywood treads and risers supported on laminated veneer lumber (LVL) stringers. LVL beams are also used at roof level to support the additional weight of the mechanical penthouse.

The interior features linear wood ceilings and acoustic wall panels, while the exterior soffits also have a linear wood finish. The exterior finishes are a combination of composite metal panels and smooth faced western red cedar siding.

项目概况

建造太平洋自闭症家庭中心（PAFC）的主要目的是将最先进的资源和研究整合到这个"知识中心"，能够更好地解决不列颠哥伦比亚省日益严峻的自闭症谱系障碍（ASD）形势。该中心将与其他较小的零星地区的设施联系起来，组成一个网络，为全省的相关个人和家庭提供学习、评估、治疗和支持服务的空间。

自闭症谱系障碍是一种影响脑部发展的障碍症。它对个人的需求、技能和能力的影响程度差异很大。根据太平洋自闭症中心基金主席Sergio Cocchia所说，在不列颠哥伦比亚省每69个新生儿就有1个自闭症谱系障碍患者，使该疾病成为儿童中最常见的神经失调症。

这栋5600平方米的三层楼建筑位于Sea Island，毗邻温哥华国际机场，两侧皆有商业建筑。该设施包括了"知识中心""信息中心""寿命研究中心"和"训练中心"，为各年龄段的客人提供了资源、教育和娱乐场所，同时也为支持他们的员工提供了行政办公和研究的场所。

在建筑的公共区域，特地设计了超大的流通和等待区，避免造成幽闭恐惧和禁闭的感觉，室内装潢简单地呈现出了一个平静的环境。"透视"手法被有策略地巧妙运用，借助外部和周围的景观帮助定位，利用内部的相邻空间（比如走廊和楼体）协助建筑内的路线导航。

木材使用

在该项目中，木材的选择符合所有设计标准，提供了较经济的大跨度结构方案，在将来能够根据自闭症研究和治疗需求的变化来重新布局。

主体结构是胶合木梁柱框架体系，地板结构是预制钉连接胶合木楼板和轻型工程木桁架搭接楼板的混合结构，两种楼板形式表面都由胶合板装饰。

胶合木柱设置在6米×6米的柱网上，以实现最大的经济性，并确保未来重新配置非承重隔墙时的灵活性。在底楼的柱截面积是2100平方毫米，楼上的柱因为承重少，截面变小。为了减少木材横纹收缩对建筑高度的影响，柱被叠加在另一个柱之上，由一个与混凝土楼面相同厚度的钢垫片来分割。这些垫片构成了连接节点的一部分，其中包括承载两侧楼板梁的底座。

钉连接胶合木板运用在公共流通区域，裸露的底面带来了巨大的视觉冲击。三个电梯井也使用钉连接胶合木板建造，楼梯的踏板和立板由旋切板胶合木（LVL）梁支撑。旋切板胶合木梁也运用在屋面上，用于支撑阁楼机房的重量。

室内采用线形木制天花板和隔声墙板，屋外悬挑部分也采用了线形木质饰面。建筑外饰面是复合金属面板和光滑的西部红柏壁板的结合。

加拿大夸扣特尔瓦加鲁斯学校
Kwakiutl Wagalus School, Canada

来源：Naturally Wood
翻译：高晓萌
校对：沈姗姗
摄影：Lubor Trubka Associates，Peter Powles Photography

Location: Port Hardy, British Columbia
Size: 1637 m²
Completion: 2016
Architects: Lubor Trubka Associates Architects
Structural engineer: CWMM Consulting Engineers Ltd.
General contractor: AFC Construction
Engineered fabrication and installation: Macdonald & Lawrence
Project owner: Kwakiutl First Nation

项目地点：哈迪港，不列颠哥伦比亚省
建筑面积：1637平方米
竣工时间：2016年
设计师：Lubor Trubka Associates Architects
结构工程师：CWMM Consulting Engineers Ltd.
总包：AFC Construction
工程木制造商及安装：Macdonald & Lawrence
项目业主：夸扣特尔瓦第一民族

Estimated environmental impact of wood use

Volume of wood products used: 361 cubic meters
U.S. and Canadian forests grow this much wood in: 1 minute
Carbon stored in the wood: 304 metric tons of CO_2
Avoided greenhouse gas emissions: 591 metric tons of CO_2
Total potential carbon benefit: 895 metric tons of CO_2

GHG emissions are equivalent to: 189 cars off the road for a year
Energy to operate 94 homes for a year

*Estimated by the Wood Carbon Calculator for Buildings, cwc.ca/carboncalculator.
*CO_2 refers to CO_2 equivalent.

"The natural cedar is the school's best feature. The foyer exemplifies the living culture of the Kwakiutl, where students learn how they will contribute to their community and world at large."

— *Marion Hunt B.S.W., Education Administrator, 99 Tsakis, Kwakiutl Band*

Project Overview

The use of wood in buildings, and as part of daily life, is an integral part of the heritage and culture of the Kwakiutl First Nation. The Kwakiutl people consider the cedar to be the tree of life, so it was only fitting that their new school would feature cedar from local forests in every aspect of the building's design.

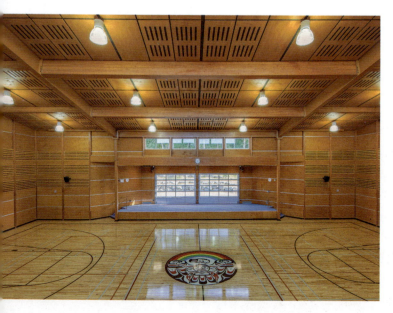

Special effort was made to highlight the use of wood as a prominent structural element as well as for interior and exterior finishes. The architects conducted multiple design planning meetings with the Kwakiutl community, which helped develop the form and plan of the school, helping guide their use of wood. The Chief and Council, community members, Elders, teachers and even future students were involved in this collaborative design process, which was critically important because all involved wanted the building to ensure the transfer of historical value.

The heart of the school – the Multi-Purpose Room – was inspired by the form and structure of a traditional Big House, featuring cedar throughout. The entry canopy, which will be erected by members of the community over time, is designed as a teaching tool representing the traditional wood framing and structure of their historical buildings.

The use of wood also offered operational benefits. To simplify and economize the construction of the large gymnasium, they used an all-wood, prefabricated wall system that could be quickly erected on site by a small crew, involving local members of the community where possible.

Wood Use

Large, round timbers are featured in the entrance foyer, while the Multi-Purpose Room consists of four western red cedar posts supporting four cedar beams. The roof structure above was framed by Douglas fir glulam purlins and I-joists. Interior walls were clad with cedar planks installed vertically, reminiscent of the traditional cladding used in Big Houses. The ceiling and surrounding corridors were finished with custom acoustic panels featuring high grade, kiln dried spruce-pine-fir turned on edge to manage sound.

The gymnasium was designed as a system of prefabricated tilt-up wood panels to speed up construction. It took just 19 days to erect the gym walls and nine days to add the roof, allowing them to enclose the gym quickly and avoid exposure to rain during construction. The panels contain conventional lumber framing between Douglas fir glulam edges. Panels were prefabricated in Mill Bay, on the southern end of Vancouver Island, and transported 450 km to the site. Once on site, the panels were raised on the slab-on-grade within five days, with the partially prefabricated wood roof erected the following week.

The gymnasium's interior wall paneling is stained plywood, chosen for its strength and durability. Custom acoustic paneling, constructed from dimension lumber, was strategically placed on the walls, with the ceiling plywood panels perforated to manage the acoustics of the big space. The gymnasium also features a maple hardwood sports floor.

Classrooms were constructed with glulam posts and beams together with conventional dimension wood framing. Wood doors and interior windows were framed by vertical grain Douglas fir architectural woodwork. Birch wood veneer finishes were also used.

使用木材预计可带来的环境影响

木材制品使用量：361 立方米
美国和加拿大森林生长出该体量的木材需要：1 分钟
木材存储碳量：304 公吨的二氧化碳
能够避免的温室效应气体排放量：591 公吨的二氧化碳
总计潜在的碳固存量：895 公吨的二氧化碳
减少的温室气体排放相当于：189 辆汽车行驶一年的排放量
支撑 94 个家庭运作一年的能量
* 由"建筑用木材的碳固存计算手册"估算，cwc.ca/carboncalculator.
* 二氧化碳指二氧化碳当量

"天然的柏树是学校最大的特色。门厅体现了夸扣特尔人的生活哲学，学生们在这里学习如何为社区，甚至整个世界作出贡献。"

—— Marion Hunt, 社会工作学学士 &
夸扣特尔理事会教育管理者

项目概况

在建筑中运用木材是夸扣特尔第一民族（Kwakiutl First Nation）日常生活的一部分，也是该民族的传统和文化中不可或缺的一部分。夸扣特尔人认为柏树是生命之树，所以在他们的新学校中采来自当地森林的柏树，这种理念体现在设计的各个方面。

设计师设计花了许多巧思来突显木材，其作为主要结构构件，同时也是内装和外装的重要饰面材料。设计师和夸扣特尔社区一同举办了多次设计会议，这有助于设计学校的框架和布局，也指导设计师如何在建筑中运用木材。夸扣特尔族族长和理事会、社区成员、老人、教师，甚至将来在学校就读的学生都参与到这种通力合作的设计过程中。这是至关重要的，因为所有的参与者都希望这栋建筑可以向人们传递出民族的历史价值。

学校的核心——多功能室的灵感出自传统"大房子"的框架和结构，从里到外突出柏树的元素。入口处的雨篷将由社区成员长期搭建，作为一种教学工具，呈现出传统木构架和民族历史建筑的结构形式。

木材的使用也提供了建造过程的便利。为了简化大型体育馆的建造并节约成本，他们运用了全木的预制墙系统，可以由一小群工人在现场迅速搭建起来，也可以让社区的成员参与其中。

木材使用

入口门厅以大的圆木为特色，多功能厅有四根西部红柏木柱支撑着四根柏木梁。上方的屋顶结构由花旗松胶合木桁梁和 I 形木龙骨组成。内墙饰面板是垂直的柏树挂板，让人想起了"大房子"里使用的传统饰面板。天花板和周围的走廊用定制的声学板装饰，这些声学板的特点是高质量，并用风干的 SPF 板在边缘包裹，以控制声音。

体育馆被设计成一个可现场安装的预制倾斜木板的结构，以加快施工速度。只用了 19 天建造体育馆的墙体，9 天安装屋顶，在施工过程中就能够迅速将体育馆封闭，避免构件在施工过程中暴露在雨水中。面板保留了花旗松胶合木板边缘的传统木边框做法。面板在温哥华岛南段的米尔湾预制的，再运到 450 千米外的工地。到了现场，在五天内完成地面基础之上的竖向板，在接下来的一周，完成部分预制的木屋顶的搭建。

体育馆的内墙挂板，因为其强度和耐久性，选择了上色的胶合板。定制的声学板由规格材制成，按一定的规则放置在墙上，同时安装天花的胶合穿孔板，以控制大空间的音效。体育馆还铺设了硬质枫木运动型地板。

教室使用胶合木梁柱，结合传统规格的木框架来建造。木门和室内窗户以垂直的花旗松木板做框，同时也采用了桦木饰面板。

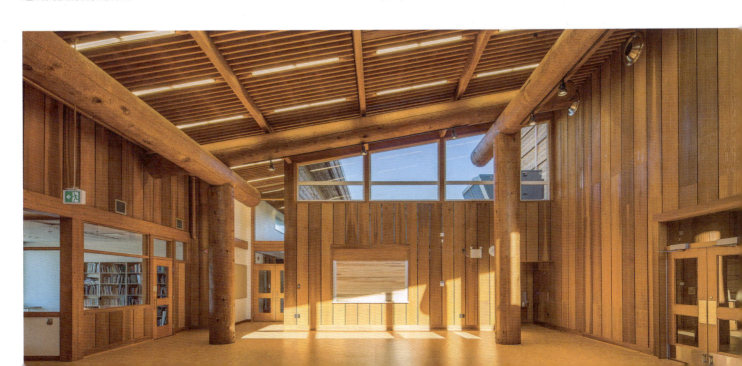

加拿大北不列颠哥伦比亚大学的木材创新研究实验室

Wood Innovation Research Lab, University of Northern B.C., Canada

来源：Naturally Wood
翻译：高晓萌
校对：沈姗姗
摄影：Michael Elkan Photography

Location: Prince George, BC
Size: 1070 m²
Completion: March 2018
Architects: Stantec
Truss Manufacturer: Winton Homes Ltd.
Structural Engineer: Aspect Structural Engineers
General Contractor: IDL Projects Inc.
Wood Supplier/Fabricator: Western Archrib
Project Partners: Government of Canada, Province of B.C., City of Prince George, University of Northern B.C.

Estimated Environmental Impact of Wood Use

Volume of wood products used: 165 cubic meters
U.S. and Canadian forests grow this much wood in: 27 seconds
Carbon stored in the wood: 144 metric tons of CO_2
Avoided greenhouse gas emissions: 307 metric tons of CO_2
Total potential carbon benefit: 451 metric tons of CO_2

GHG Emissions Are Equivalent to: 95 cars off the road for a year
Energy to operate 48 homes for a year

*Estimated by the Wood Carbon Calculator for Buildings, cwc.ca/carboncalculator.
*CO_2 refers to CO_2 equivalent.

"This building has caught the attention of Passive House researchers around the world because it demonstrates how an industrial structure, constructed with wood, in a northern climate exceeds a rigorous, internationally recognized energy efficiency standard."

—— *Dr. Guido Wimmers, Chair/Associate Professor, Engineering Graduate Program, UNBC*

项目地点：乔治王子城, 不列颠哥伦比亚省
建筑面积：1070 平方米
竣工时间：2018 年 3 月
设计师：Stantec
桁架生产商：Winton Homes Ltd.
结构工程师：Aspect Structural Engineers
总包：IDL Projects Inc.
木材供应商：Western Archrib
项目合作伙伴：加拿大政府，不列颠哥伦比亚省，乔治王子城，北不列颠哥伦比亚大学

使用木材预计可带来的环境影响

木材制品使用量：165 立方米
美国和加拿大森林生长出该体量的木材需要：27 秒
木材存储碳量：144 公吨的二氧化碳
能够避免的温室效应气体排放量：307 公吨的二氧化碳
总计潜在的碳固存量：451 公吨的二氧化碳
减少的温室气体排放相当于：95 辆汽车行驶一年的排放量
支撑 48 个家庭运作一年的能量
* 由"建筑用木材的碳固存计算手册"估算, cwc.ca/carboncalculator.
* 二氧化碳指二氧化碳当量

"这栋建筑吸引了全球被动屋研究者的关注，因为它展示了在北方寒冷地区，用木材建造的工业建筑如何达到甚至超过严格的、国际认可的能效标准。"

—— Guido Wimmers 博士
北不列颠哥伦比亚大学工程研究生课程 首席 / 副教授

Project Overview

The Wood Innovation Research Lab (WIRL) at the University of Northern British Columbia (UNBC) gives students and researchers much needed space to test state-of-the-art building systems. The WIRL provides an opportunity to study ways to integrate wood into more structural designs for industrial buildings.

The research facility was built using some of the same innovative wood building products and systems that students learn about in the lab. The building is located adjacent to the Wood Innovation and Design Centre, which houses UNBC's Master of Engineering in Integrated Wood Design program.

The WIRL is noteworthy in that it is the first industrial building in North America certified to rigorous Passive House energy standards. Certified Passive House buildings use up to 90 percent less energy for heating and cooling and up to 70 percent less energy overall compared with standard buildings. WIRL's achievement of Passive House standard is particularly remarkable given the large volume-to-floor-area ratio of the structure and the cold climate of its location in northern British Columbia.

The project team integrated strong floor and wall structures to accommodate heavy testing equipment; they also designed a superstructure which supports the roof and the overhead crane used to maneuver heavy materials. All building structural systems were constructed of wood.

The WIRL is a showcase for innovative wood construction and pioneers the application of wood with high performance design standards in industrial buildings.

Wood Use

The WIRL is a single-story mass timber structure, composed of glue laminated timber (glulam) columns and beams on a concrete raft slab foundation. The building is 10-metres in height and consists of high-head lab space, classrooms and office space.

Instead of using standard wood studs, the 10-metre tall wall panels are framed with prefabricated 0.5-metre thick upright wood trusses. The walls are insulated with mineral wool specifically designed to achieve the high thermal performance required for Passive House certification. The trusses were fabricated by a Prince George-based company using dimension lumber from a local sawmill.

Designers also used I-joists for the second-level floors. Sheet goods used to sheathe the floors, roof and wall assemblies were left exposed to provide the interior finish for the lab portion of the building.

Researchers conducted a comparative Life Cycle Assessment on the WIRL so that the team could quantify the relative impact of the wood material selection compared to the impact of the operational energy of the building. By lowering the operating energy requirements through Passive House design, global warming impact of the WIRL structure was reduced by 70 percent, primarily due to the reduction in energy used for heating.

Renewable biogas is used to heat the 1070 square-meter structure. Due to the high performance of the Passive House design and wood use in the structure, the low heating requirement for the building is similar to that of a typical family home. The decision to use wood, when considering the environmental impact from materials alone, showed a 22 percent improvement over steel due to wood's lower carbon footprint and ability to sequester carbon.

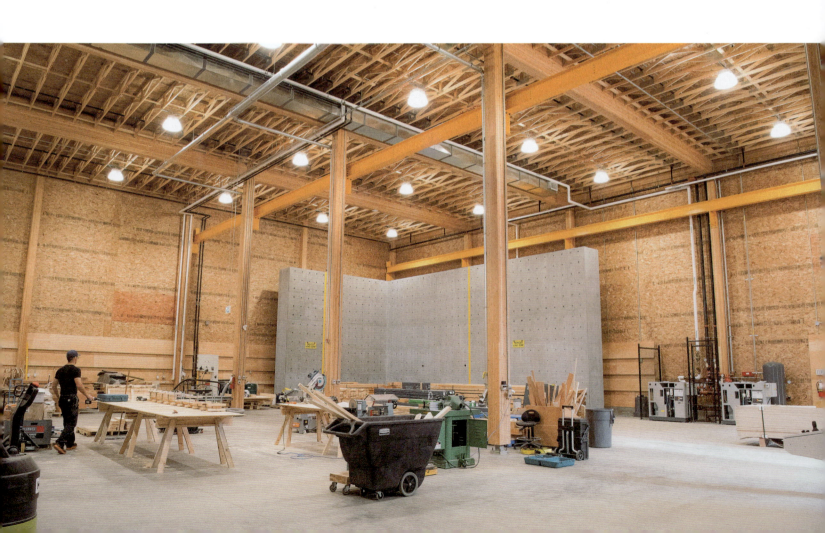

项目概况

位于北不列颠哥伦比亚大学的木材创新研究实验室为学生和研究者提供了足够的空间考察最顶尖的建构系统。木材创新研究实验室为研究如何将木材更多地融合到工业建筑的结构设计中提供了机会。

该研究设施是用一些学生在实验室学习到的相同的创新木结构产品和系统建造的。该建筑毗邻学校的木材创新设计中心,该中心是北不列颠哥伦比亚大学综合木材设计工程硕士项目的所在地。

值得注意的是,木材创新研究实验室是北美第一个通过严格的被动房能源标准认证的工业建筑。与普通建筑相比,经过认证的被动房建筑在供暖和制冷方面的能耗减少了90%,整体能耗减少了70%。木材创新研究实验室达到被动房标准,这个成就是很大的,因为它的建筑容积率很大,并且该建筑位于寒冷的不列颠哥伦比亚省北部。

项目团队整合了强度高的楼面和墙体结构,足以承受重型检测设备;他们还设计了一个上部构造,用来支撑屋面和操控重型材料的吊车。整个建筑的结构体系都是用木材建造的。

木材创新研究实验室对创新木结构的一种展示,也是运用木材建造高性能设计标准的工业建筑的先锋。

木材使用

木材创新研究实验室是单层的重木结构,在混凝土的基础之上建造胶合木梁柱。该建筑高10米,由挑高的实验室、教室和办公空间组成。

10米高的墙板没有采用标准的木龙骨,而是采用了0.5米厚的预制木桁架做框架。为了使墙体能达到被动房认证所需的热工性能,墙体保温采用了矿棉。桁架由乔治王子城的公司预制,使用了当地锯木厂的规格材。

设计师们使用了I型木龙骨来搭建二层楼板。地面以薄板覆盖,屋顶和墙体的装配结构件暴露在外,为建筑的实验室部分提供了室内装饰。

研究人员对木材创新研究实验室进行了全生命周期评估,以便团队能够量化木材的环境影响,其与建筑运行能耗的环境影响做对比。根据被动房设计标准设计的木材创新研究实验室,通过降低其建筑运行能耗,其结构形式对全球变暖的影响降低了70%,主要是因为供暖能耗减少了。

这栋1700平方米的建筑使用可再生沼气来供暖。由于被动房设计和结构用木材的高性能,该建筑的较低的供暖要求,与一个典型家庭住宅建筑类似。考虑到材料本身对环境的影响,由于木材的低碳足迹和固碳能力,使用木材可以比使用钢铁优化22%。

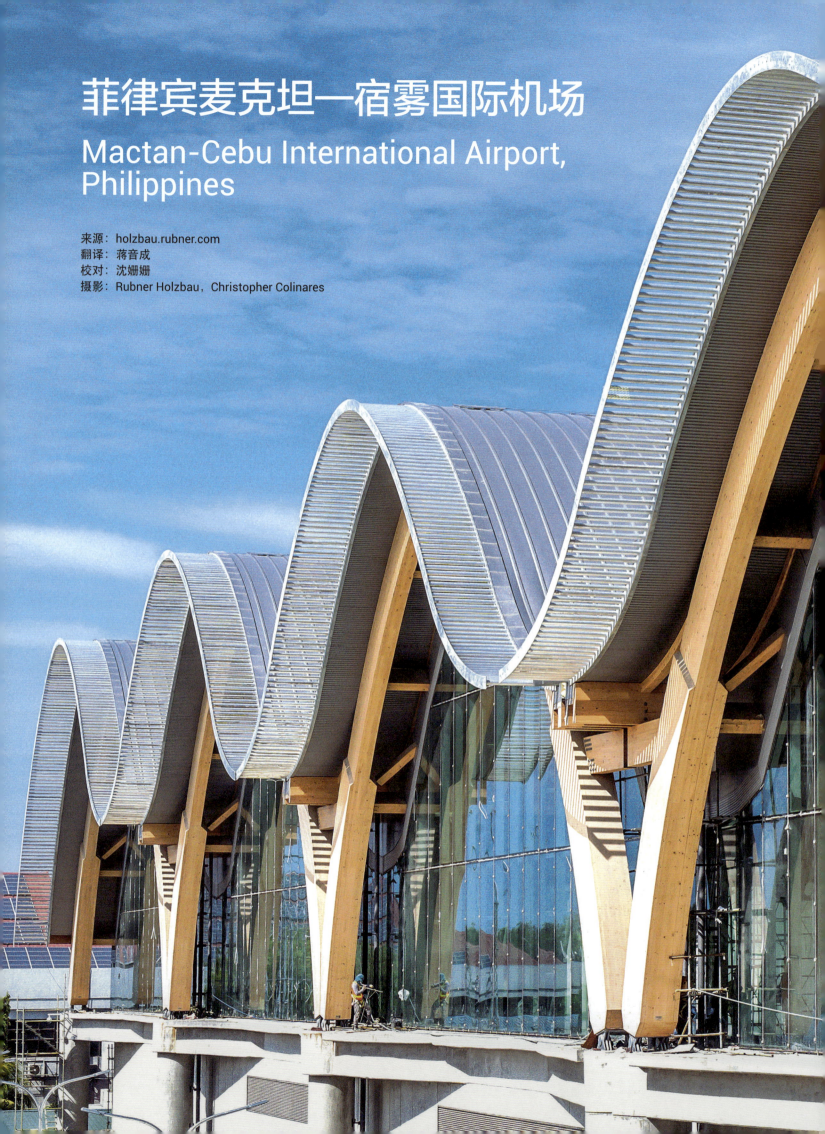

菲律宾麦克坦—宿雾国际机场
Mactan-Cebu International Airport, Philippines

来源：holzbau.rubner.com
翻译：蒋音成
校对：沈姗姗
摄影：Rubner Holzbau，Christopher Colinares

Asia now has its first airport with a roof and frame made entirely from glulam, with Mactan Cebu International Airport, the second largest in the Philippines, gaining a new terminal. The aim was for the visitors to enjoy a warm welcome and an equally warm departure in a resort-like ambience.

亚洲现在有了第一个屋顶和结构框架完全使用胶合木的机场。麦克坦—宿雾国际机场，菲律宾第二大机场，建造了一个新的航站楼。目标是让旅客感受到热烈的欢迎，在离开的时候沐浴着同样温暖的度假气氛。

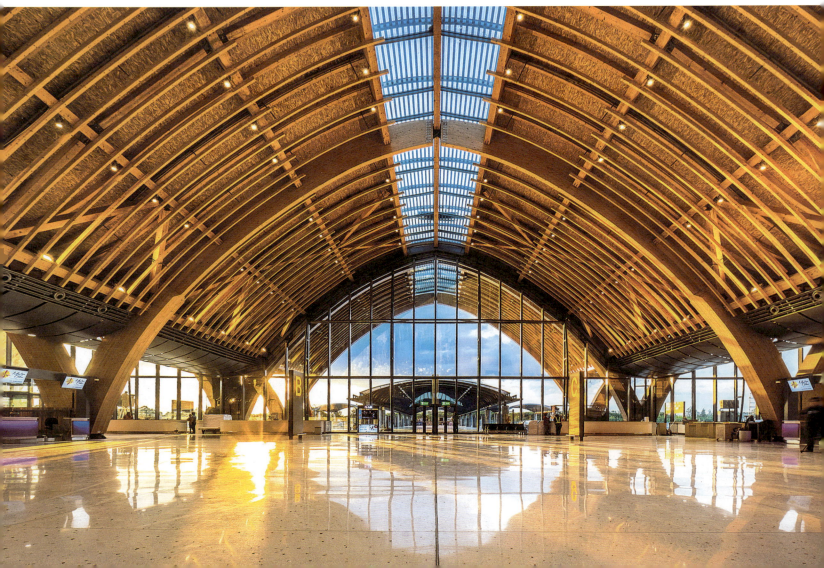

Light architecture combined with the right material was required to reflect the friendly and open culture of the Philippines. For both ecological and traditional reasons, wood was the final choice. Before construction could begin, the structural engineers at Rubner Holzbau built a full-scale test model. This enabled them to make sure that the joints were able to handle any seismic movements and that the roof beams anchored in the ground floor's concrete posts stabilise the building in this hurricane and earthquake-prone part of the world.

轻盈的建筑结合有轻盈感的材料呈现出菲律宾友好而开放的文化传统。基于生态和传统原因，木材成为最终的选择。在施工开始之前，Rubner Holzbau 的结构工程师建造了一个 1：1 的测试模型。这个模型协助设计师，以确保连接节点能够承受任何地震活动荷载，确保固定在地面层混凝土柱上的屋面梁可以在飓风和地震高发地区保持稳定状态。

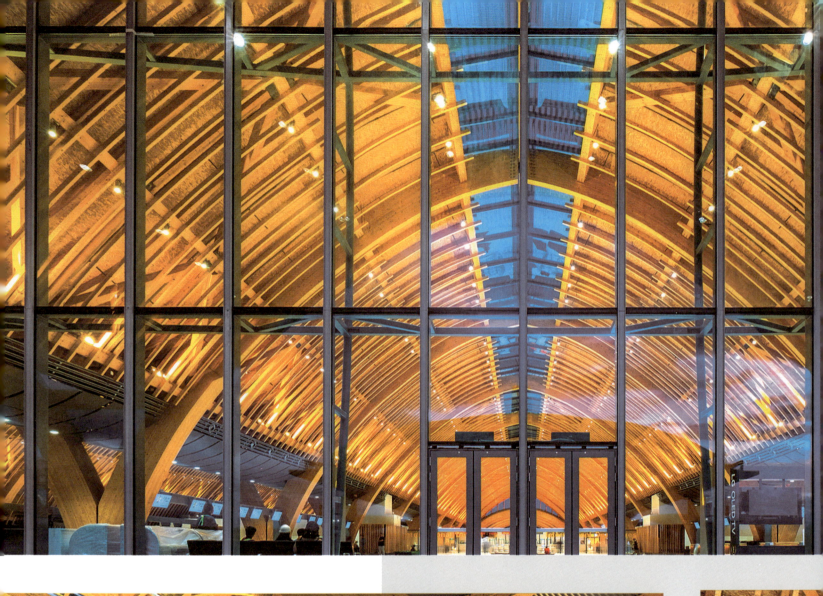

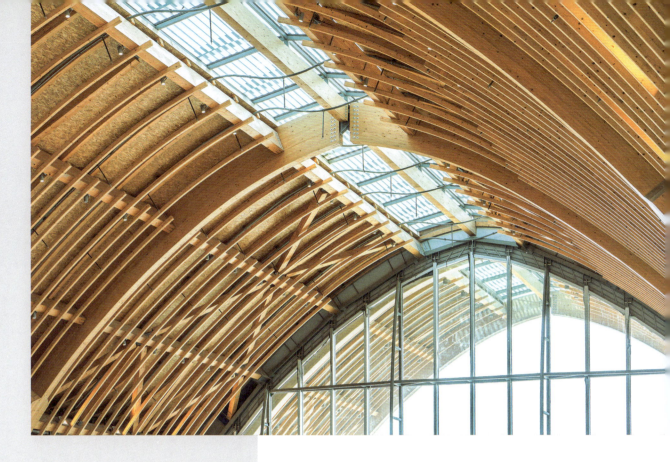

In total, 4500 cubic metres of glulam has been used for the vaulted structure, which has a height of 15 metres and a span of 30 metres.

这栋穹顶建筑总共使用了 4500 立方米的胶合木。这栋建筑跨度为 30 米,高为 15 米。

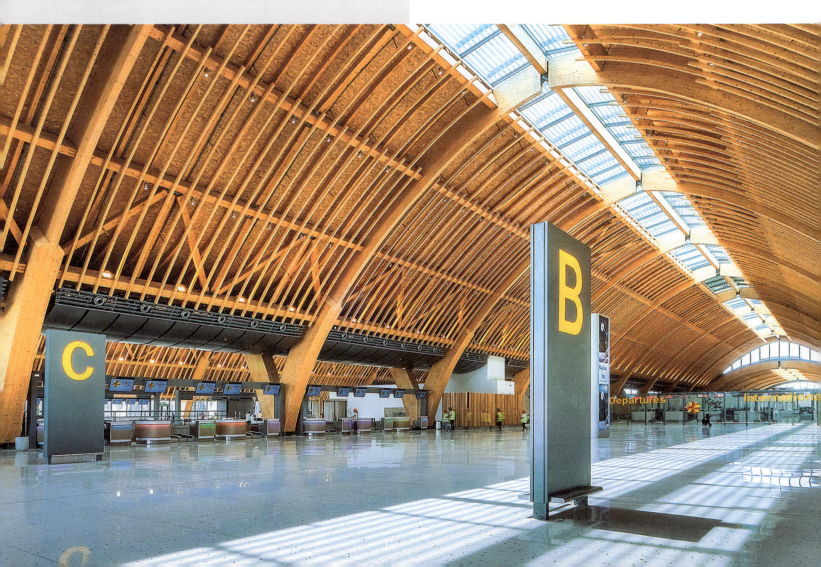

挪威奥斯陆机场第二航站楼
Oslo Airport, Terminal 2 in Oslo, Norway

作者：Erik Bredhe
翻译：蒋音成
校对：沈姗姗
摄影：Ivan Brodey，Dag Spant，
Nordic - Office of Architecture

Nordic - Office of Architecture was founded in 1979 and now has 144 employees. Its head office is in Oslo, with other offices in Copenhagen and London. Nordic designs everything from hospitals and schools to large housing developments. Over the years, they have become specialists in airports, with examples including Istanbul New Airport.

Client: Avinor Oslo Lufthavn
Structural engineer: Aas-Jacobsen
Cost: NOK 14 billion
Size: 115000 square metres (newbuild)

Natural material creates an atmospheric environment and reduces emissions at the world's greenest airport.

Dreams and Anticipation, Queues and Delays. Most of us are conflicted about airports. Andy Warhol was one of their greatest advocates, as they provided his favorite foods, favorite toilets and favorite peppermint pastilles. British design critic Reyner Banham, on the other hand, has expressed his disdain and called them a "demented amoeba". Christian Henriksen, partner and design manager at Norway's Nordic Office of Architecture, has a more nuanced relationship with the buildings.

"Working with airports is fascinating. Many of them look the same the world over, so there are great opportunities to do something new, special and different. However, everyone knows that aircraft are responsible for considerable air pollution. As an architect, I therefore want to do everything I can to reduce the carbon footprint of the airport itself."

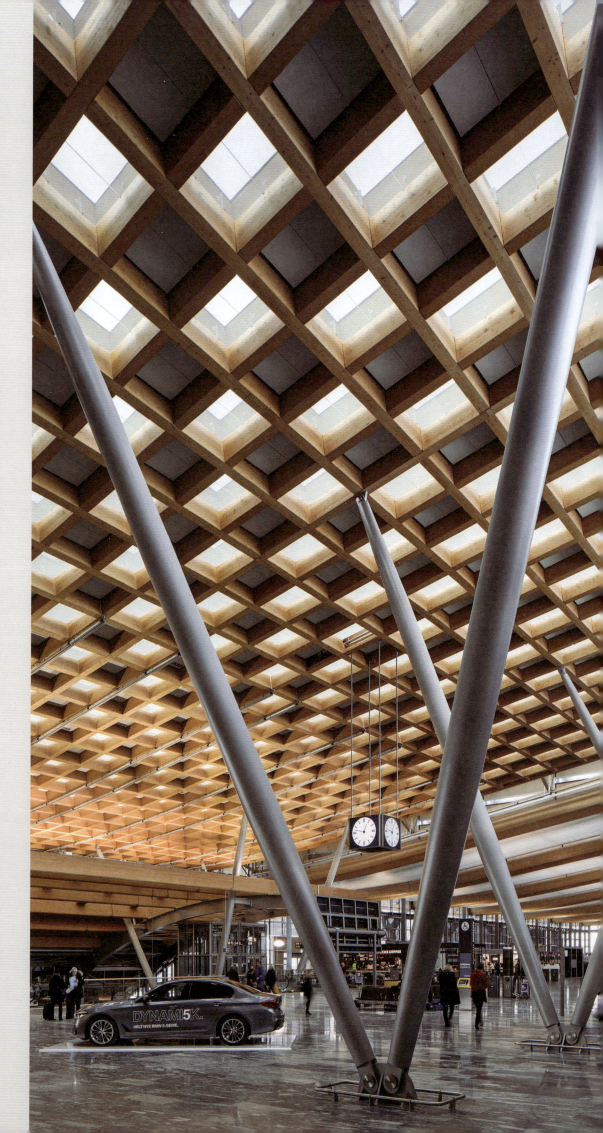

挪威北欧建筑事务所成立于1979年，现在有144名员工。它的总部在奥斯陆，在哥本哈根和伦敦也有办公室。挪威北欧建筑事务所的业务范围包括医院和学校到大型住宅发展项目。经过多年沉淀，他们变成了机场设计专家，其成功的案例包括伊斯坦布尔新机场。

客户：Avinor Oslo Lufthavn
结构工程师：Aas-Jacobsen
花费：1400万挪威币
规模：115000平方米（新建）

自然材料营造了一个有气氛的环境，也为世界最绿色机场降低了二氧化碳排放。

梦想和期待，排队和延误。我们大多数人都对机场感到抵触。Andy Warhol是他们最大的支持者之一，因为他们提供了他最喜欢的食物、最喜欢的厕所和最喜欢的薄荷馅饼。另一方面，英国设计评论家Reyner Banham表达了他的蔑视，称它们为"疯狂的变形虫"。挪威北欧建筑事务所合伙人兼设计经理Christian Henriksen与这些建筑有一种更为微妙的关系。

"参与机场项目很有意思。世界各地的许多机场看起来都一样，所以我们就有很多机会去做一些新的、特别的和不同的事情。然而，每个人都知道飞机会造成相当大的空气污染。因此，作为一名建筑师，我希望尽我所能减少飞机场本身的碳排放。"

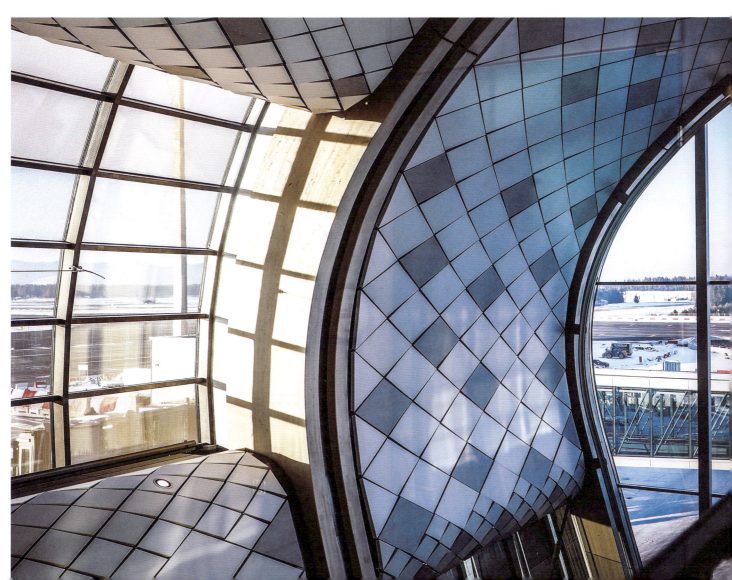

Christian Henriksen has been involved in designing airports all over the globe, including in Turkey, India, France, Iceland, Uganda and Sweden. Nordic also designed Oslo Airport, which was completed in 1997. Then last spring – 20 years later – an airport extension, also by Nordic, was opened. The airport was originally built to handle 17 million passengers a year, but now it can handle an impressive 32 million.

"It was a huge job that took a great deal of planning. Everything you do in the construction phase has to be reported well in advance so that air traffic and safety are not affected in any way. What's more, the safety measures for airports have changed a lot in recent years, which meant we were forced to keep the design of certain parts open and see how things developed." says Christian Henriksen.

Keeping the airport working seamlessly during the extension work was not easy – but it was essential. As well as being at the heart of Norway's expansive air traffic system, Oslo Airport is also what finances the operation of many of the 60 or so smaller airports across this topographically challenging country. But the project went well, with all the planning paying off. During the construction phase, Oslo Airport was named the most punctual airport in Europe – twice.

One of the watchwords for the project was simplicity. Travellers want to feel relaxed, safe and well looked after. The old buildings set the tone for the extension, as it was important for them not to feel like two separate units. The flow through the airport was always a priority. Thanks to its logical layout, with several recurring elements, the building is more or less self-explanatory. As such, it requires less signage and fewer PA announcements. The pervading sense of calm and the striking architecture are intended to make arriving passengers feel welcome to Norway and Scandinavia.

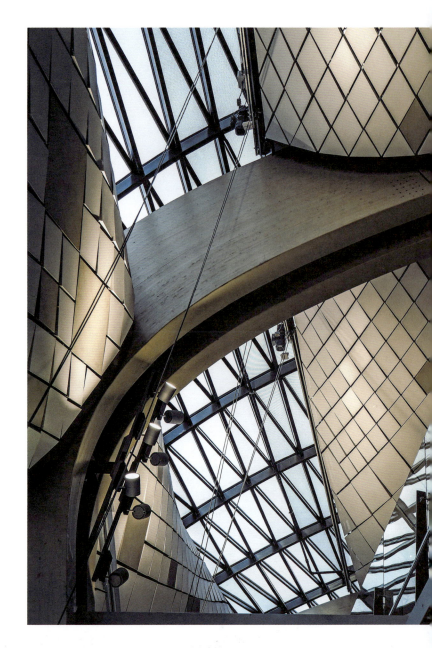

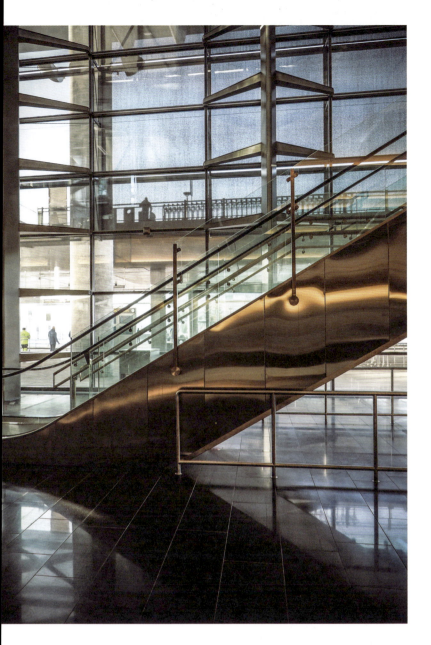

Christian Henriksen 曾参与了全球范围内的机场设计项目，包括土耳其、印度、法国、冰岛、乌干达和瑞典的机场。挪威北欧建筑事务所也参与设计了奥斯陆机场，它于1997年完工。在20年后的春天，同样由挪威北欧建筑事务所 (Nordic) 设计的机场扩建项目正式启用。这座机场最初设计的每年接待人数为1700万名乘客，但现在它可以接待3200万名乘客。

"这是一个需要花大力气去计划的巨大工作。你在施工阶段所做的每件事都必须提前做好报告，这样航空交通和安全才不会受到任何影响。更重要的是，近年来机场的安全措施发生了很大变化，这意味着我们需要在某些部件的设计上留出空白，并观察事态的发展。"Christian Henriksen 说。

在扩建工作期间，保持机场的无缝运行并不容易，但这是至关重要的。奥斯陆机场不仅是挪威庞大的空中交通系统的核心，也是这个地形多变国家的60多个小型机场运营的资金来源。该项目进展顺利，所有的规划都得以实现。在建设阶段，奥斯陆机场两次被评为欧洲最准点的机场。

该项目的口号之一是简单。旅行者想要放松、安全，得到良好的照顾。旧建筑为扩建定下了基调，因为重要的是使它们不像两个独立的单元。机场的流线一直是头等大事。由于它的逻辑布局，有几个重复的元素，建筑或多或少是不言自明的。因此，它可采用更少的标识和更少的 PA 公告。无处不在的平静感和引人注目的建筑旨在让到达的乘客感到在挪威和斯堪的纳维亚是受欢迎的。

"We focused a great deal on the dearly held values of Scandinavia and let them manifest themselves in both the forms and materials. The new terminal building thus has a sense of openness and security. Transparency and clarity. And closeness to nature." says Christian Henriksen.

Large panoramic windows enable passengers to enjoy the landscape around the airport. Many places in the terminal have walls of plants and small water features that recall the natural beauty of Norway. All the materials were meant to be used in as natural a form as possible. If it's glass, it should look like glass. Stone should be stone. And wood, which plays a starring role at Oslo Airport, should take pride of place. The architects used wood in two main ways: structurally and for atmosphere.

"We wanted to use wood in the structure and keep it exposed. Wood defines us here in Scandinavia and it would therefore also define Oslo Airport. We've used wood in various different parts of the airport where we wanted to create a particular atmosphere- in the areas where the passengers spend a lot of time or where they want a certain amount of calm." states Christian Henriksen.

"我们非常重视斯堪的纳维亚人珍视的价值观,并让它们在形式和材料上得到自我体现。因此,新的航站楼具有开放和安全的感觉,透明而清晰,亲近自然。"Christian Henriksen 说。

巨大的全景窗使乘客可以欣赏到机场周围的景色。航站楼的许多地方都有植物墙和小水景,让人想起挪威的自然美景。所有的材料都被设计成尽可能自然的形式。如果是玻璃,它应该看起来像玻璃,石头应该是石头。木材在奥斯陆机场扮演着重要角色,它应该占据主导地位。建筑师主要以两种方式来使用木材:结构作用和烘托氛围。

"我们想在结构中使用木材,并将其暴露在外。木材在斯堪的纳维亚定义了我们,因此它也定义了奥斯陆机场。我们在机场的各个不同部分都使用了木材,我们想在这些地方创造一种特别的氛围。乘客会花费大量时间待在这些区域里,或者他们需要待在一定程度平静的地方。"Christian Henriksen 说。

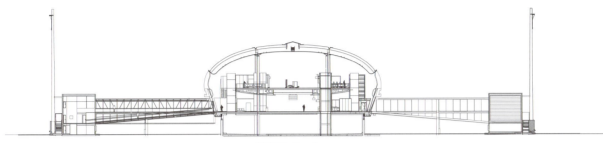

剖面图 Section

Nordic expanded the existing terminal and modernised the train station that serves it. But the biggest change was the construction of a 300-metre long pier that extends out onto the tarmac airside. The slightly futuristic, rounded pier is held up by a series of glulam arches that use wood from Scandinavian forests.

Where the pier meets the original terminal, the rounded form spreads out like butterfly wings to bring the two buildings together in a natural way. This area houses the duty-free stores and is where the domestic and international travellers part ways – domestic upstairs, international downstairs.

The façade is mostly glass to let in as much daylight as possible, while the roof is finished with oak cladding. Around 20 centimetres beneath this is an additional protective roof that deals with water run-off and keeps the external wooden façade well ventilated. A transparent and UV-resistant paint is the only treatment used to protect the oak cladding. The wooden façade gives the pier an attractive aesthetic, but its most important function relates to airport safety.

"A pier like this that extends out towards the runways can cause problems for the air traffic control tower. The signals they send to the aircraft on the ground are highly sensitive to distortion and can easily be reflected, for example by metal, which creates false signals that can lead the plane astray. But with a wooden façade, the signals die when they hit the pier." says Christian Henriksen.

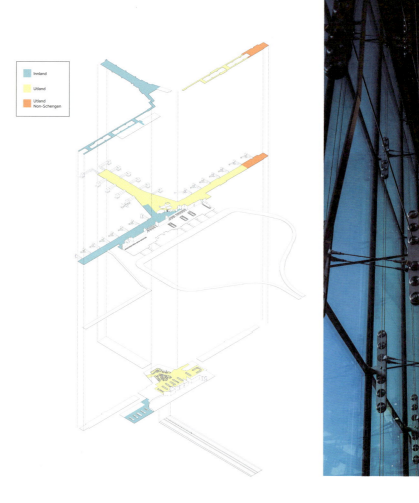

挪威北欧建筑事务所设计了现有航站楼的扩建，并对为其提供服务的火车站进行了现代化改造。但最大的变化是建造了一个300米长的航站码头（楼），延伸到跑道。这个有点未来感的圆形航站码头（楼）由一系列胶合木拱支撑，这些拱采用了斯堪的纳维亚森林的木材。

在新航站码头（楼）与原始航站楼的交汇处，圆形的形式像蝴蝶翅膀一样展开，以自然的方式将两座建筑连接在一起。这个区域是免税商店的所在地，也是国内和国际游客分开的地方——国内旅客上楼，国际旅客下楼。

立面主要是玻璃，让尽可能多的阳光进入，而屋顶则是橡木覆层。大约20厘米以下是一个额外的保护屋顶层，用于处理雨水径流，保持外部木质立面有良好的通风。一种透明的、抗紫外线的涂料是用于保护橡木覆层的唯一处理方法。木质外立面材质赋予航站码头（楼）迷人的美感，其最重要的功能与机场安全因素有关。

"像这样延伸到跑道的航站码头会给空中交通管制塔带来麻烦。它们向地面上的飞机发送的信号非常容易被扭曲和反射，比如金属。这样会产生错误信号，导致飞机误入歧途。但由于采用了木质外立面时，当信号发射到航站码头时，信号就会消失，" Christian Henriksen 表示。

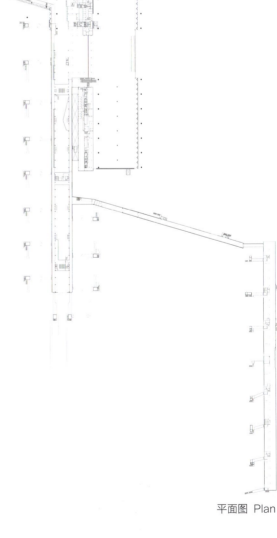

平面图 Plan

立面图 Elevation

Such wide use of wood was also a way to meet the strict environmental requirements set by the clients. All the materials were carefully chosen based on their carbon footprint. The concrete in the design was mixed with volcanic ash and more than half of the steel is recycled metal. The pier has its rounded shape to minimize its contact surface and so save energy. During the winter, the snow cleared from the runways is stored in a space below the terminal so that it can then be used to cool the terminal during the summer months. This reduces the terminal's energy consumption by 2 GWh per year. As a result, Oslo Airport is ranked as "excellent" under the BREEAM environmental certification scheme and the extension is considered one of the world's greenest terminals: the airport has cut its carbon emissions by 35 percent and halved its energy consumption compared with before the extension – while at the same time doubling the airport's capacity.

Christian Henriksen looks forward to using wood to a greater extent in even more Scandinavian projects. But he does see a cloud on the horizon.

"We have the knowledge and the material, and there are highly sophisticated techniques for using wood in large structures. But certain conditions and regulations are impeding progress, which means we don't always manage to unlock the huge potential. That's something we really have to change," he says.

如此广泛的使用木材也是满足客户对严格的环境保护要求的一种方式。所有材料都是根据它们的碳排放精心挑选的。在设计中，混凝土是与火山灰混合过的，超过一半的钢材是回收金属。航站码头呈圆形的形状，以尽量减少其接触面，从而节省能源。在冬天，跑道上的积雪被储存在候机楼下面的一个空间里，这样在夏天的几个月里就可以用来给候机楼降温了。这将使航站楼的能源消耗每年减少2千瓦时。因此，奥斯陆机场被 BREEAM 环境认证机构列为"优秀"项目，同时也被认为是世界上最绿色的航站楼之一：与航站楼扩建之前相比，机场已经减少了35%的碳排放和减半了其能源消耗；与此同时，机场的容量翻了1倍。

Christian Henriksen 期待在斯堪的纳维亚地区的项目中可以更多地使用木材。但他确实也看到了问题。他说："我们有知识和材料，在大型建筑中使用木材是非常复杂的技术。但某些条件和规定阻碍了其进步，这意味着我们并不总是能够释放出巨大的潜力。这是我们必须改变的。"

上海西郊宾馆意境园多功能厅

Yijingyuan Multi-function Hall in Xijiao State Guest Hotel, Shanghai

作者：何如
翻译：付维舟，陶亮
校对：东鸿，蒋音成
摄影：苏州昆仑绿建木结构科技股份有限公司（Suzhou Crownhomes Co., Ltd.）

Project Name: Yijingyuan Multi-function Hall
Location: Xijiao State Guest Hotel, Shanghai
Floor area: 856.64m²

Architectural design: Green-A Architecture & Decoration Design Co., Ltd.
Project Leader: Weiwei Zhu

Architect: Ru He (Principal Architect), Fenglong Tang, Sha Zhu, Ying Gao
Structural Engineer: Weizhou Fu (Canada Wood), Xiaotian Xing (Canada Wood)
Timber Structure Consultant: Jing Kong (Equilibrium Consulting)
General Contractor: Suzhou Crownhomes Co., Ltd.
Engineered Wood Fabricator: Suzhou Crownhomes Co., Ltd.
Developer: Shanghai Donghu Group Co., Ltd.
Completion: October, 2018

项目名称：意境园多功能厅
项目地点：上海西郊宾馆
建筑面积：856.64 平方米

设计单位：上海绿建建筑装饰设计有限公司
设计总负责人：朱蔚蔚

建筑设计：何如，汤凤龙，朱沙，高英
结构设计：付维舟（加拿大木业），邢笑天（加拿大木业）
结构顾问：孔晶（Equilibrium Consulting）
工程总包：苏州昆仑绿建木结构科技股份有限公司
工程木供应商：苏州昆仑绿建木结构科技股份有限公司
开发商：上海市东湖（集团）有限公司
竣工时间：2018 年 10 月

Yi Jing Yuan multi-function hall was built in the Xijiao State Guest Hotel in Shanghai, which occupies a broad, beautiful landscape with lush trees. Since its completion in 1960, as Shanghai's largest and highest ranking of the National Guest House, Xijiao State Guest Hotel has received many domestic and foreign dignitaries and held many important events, making its high.status in the contemporary development of Shanghai.

西郊宾馆意境园多功能厅建于上海西郊宾馆内部，该宾馆占地广阔、环境优美、树木繁茂，自 1960 年建成以来即作为上海规模最大、等级最高的国家迎宾馆，曾接待诸多国内外政要，举办若干重大事件与活动，在上海当代城市发展中具有重要的地位。

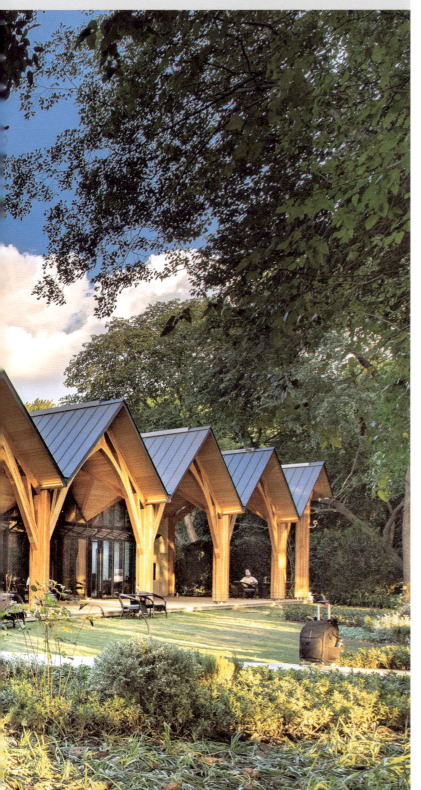

In 2016, the Ministry of Natural Resources of Canada, the BC Forestry Department of Canada and the Donghu Group who owns the Xijiao Hotel reached an agreement on further encouraging the development of wood structure, and chose the planned Yi Jing Yuan multi-function hall as a practical case, for the application and promotion of wood structure building. Canada Wood worked together with Donghu Group on the site selection, discussion and investigation, and recommended Green-A Architecture & Decoration Design Co., Ltd. as the architectural design company of the project. Canada Wood also provided the structural design and on-site supervision.

2016年，加拿大自然资源部、卑诗省林业厅与西郊宾馆所属的东湖集团关于进一步推动木结构发展达成协议，并将规划中的西郊宾馆意境园多功能厅作为实践案例，尝试木结构建筑的建造、应用与推广。经由加拿大木业协会与东湖集团前期选址、讨论及调研，推荐上海绿建建筑装饰设计有限公司负责该项目建筑设计，并由加拿大木业协会提供结构设计、现场监督等多方面支持。

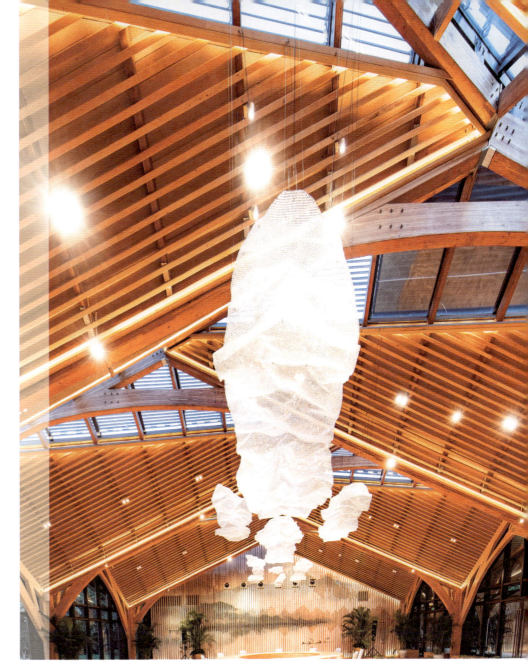

The building area of Yi Jing Yuan Multi-function hall is about 857 square meters. Located on the east side of the lake embraced by dense trees, the building with elegant and private setting offers the hotel guests with dining, meeting and event space. The project is a single-storey glulam wood structure building, with a modest and humble profile, while adopting a delicate and rich interior space, makes it a perfect fit into the surrounding. The offsite prefabrication of wood structure components allows the onsite construction to be completed in an efficient and fast manner and caused minimum damage to the site land and surrounding landscapes. By using Wood, the ecological materials, the natural, green and sustainable development of the concept of construction has really been realized.

平面图 Plan

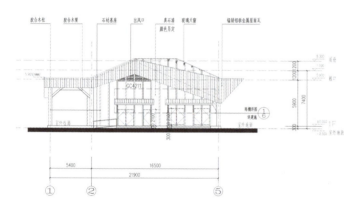

立面图 Elevation

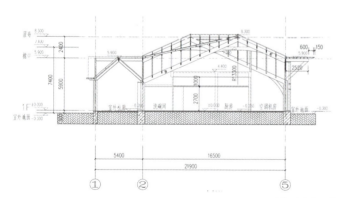

剖面图 Section

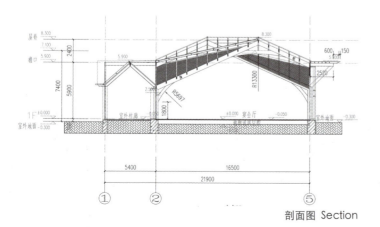

剖面图 Section

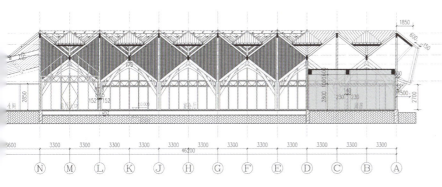

立面图 Elevation

西郊宾馆意境园多功能厅总建筑面积约857平方米。项目选址位于西郊宾馆内一座湖畔东侧，场地四周原有大量高大树木及小山丘，景观幽雅静谧。项目采用单层胶合木结构建筑，形态谦逊低调，空间细腻丰富，与所处的环境充分契合。木结构及其施工方式的选择，使得整个建造过程尽量减少对场地的侵扰与破坏，最大限度地保留了原有的树木及景观氛围。木材这种生态材料的运用，使自然、绿色、可持续发展的建造理念真正得以实现。

The architectural form simulates the attitude of tree growth. The structure columns and the triangular roof frame are naturally handed over and integrated, and the columns are staggered to form a three-dimensional spatial structure, which make the space more charming and interesting. The roof shape is composed by several equal-size folded plates staggered with each other, in clear geometric logic, to create a concise and dynamic indoor space, with a well-organized order, while providing the traditional simplicity and elegant charm. Skylights are set on the top of the folded plates to improve the lighting conditions of the forest buildings. When the sun shines through the skylights, it seems that the branches and leaves of the trees cast halos.

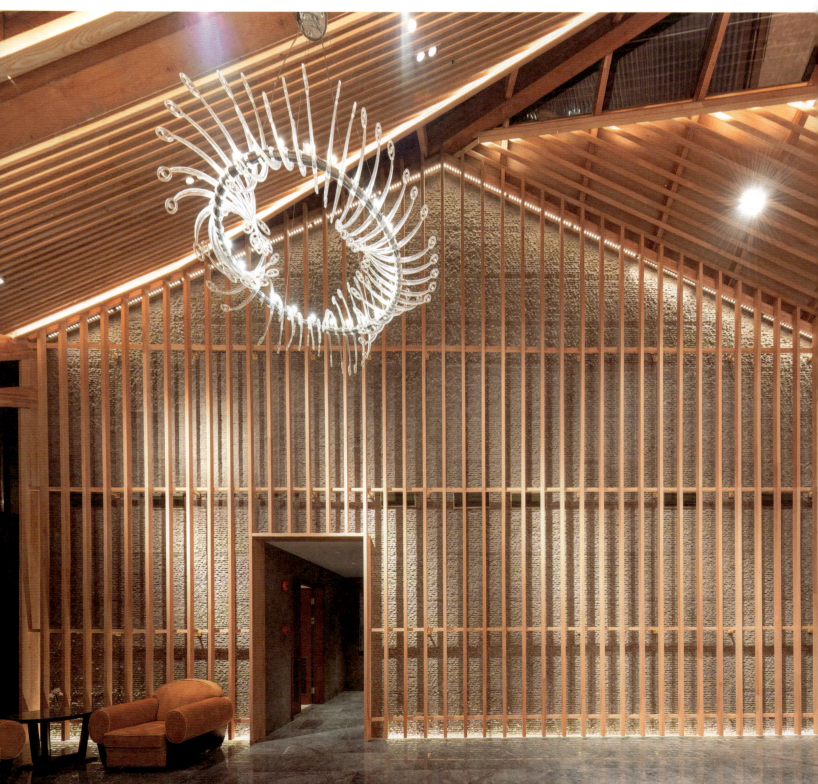

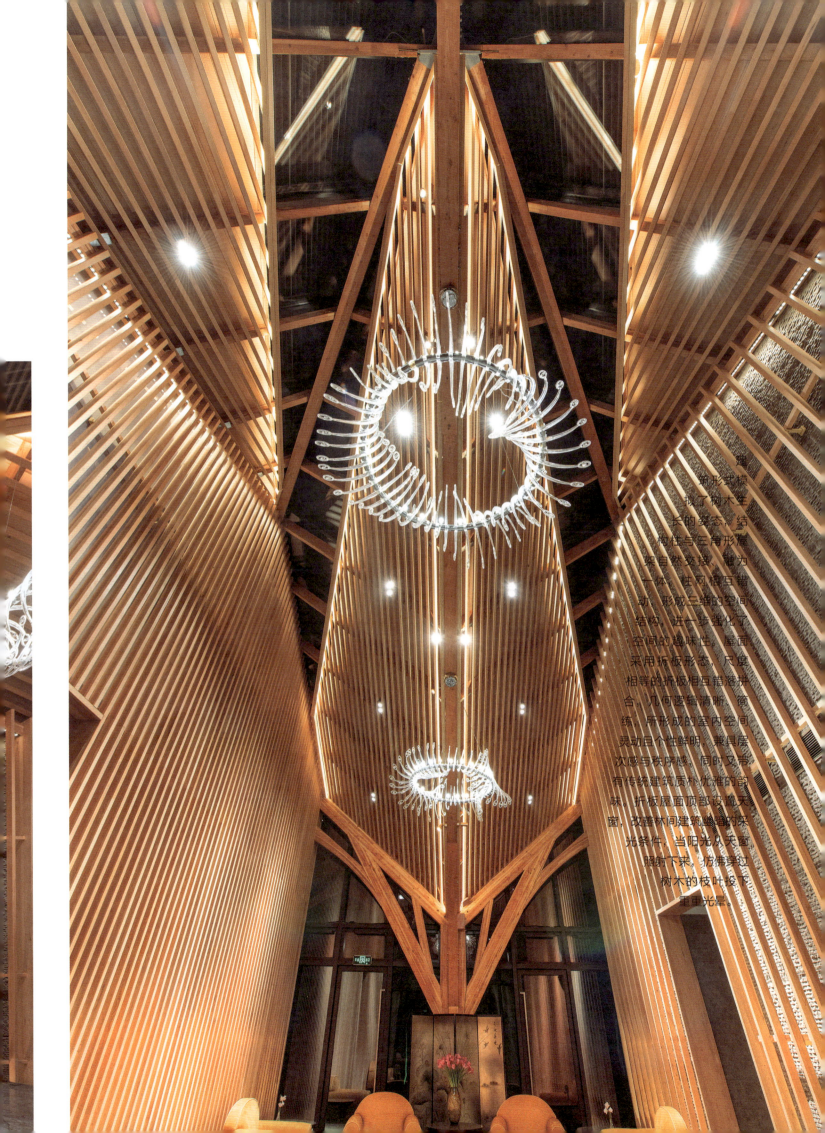

筑形式模拟了树木生长的姿态，结构性与三角形屋架自然交接，融为一体，柱网相互错动，形成三维的空间结构，进一步强化了空间的趣味性。屋面采用折板形态，尺度相等的折板相互错落拼合，几何逻辑清晰，简练。所形成的室内空间灵动且个性鲜明，兼具层次感与秩序感，同时又带有传统建筑质朴优雅的韵味。折板屋面顶部设置天窗，改善林间建筑幽暗的采光条件。当阳光从天窗映射下来，仿佛穿过树木的枝叶投下重重光晕。

With a limited construction budget, the building has created a wonderful space and fascinating form, fully reflecting the natural, warm texture of the wood structure, as well as the aesthetic sense of geometry and mechanics.

The interior finishing with wood reinforces the touch of natural beauty and design versatility. The wood grille backdrop represents the natural imagery of mountains that has long been read as a metaphor for the individual's values and beliefs of unity of nature and human in Chinese philosophy.

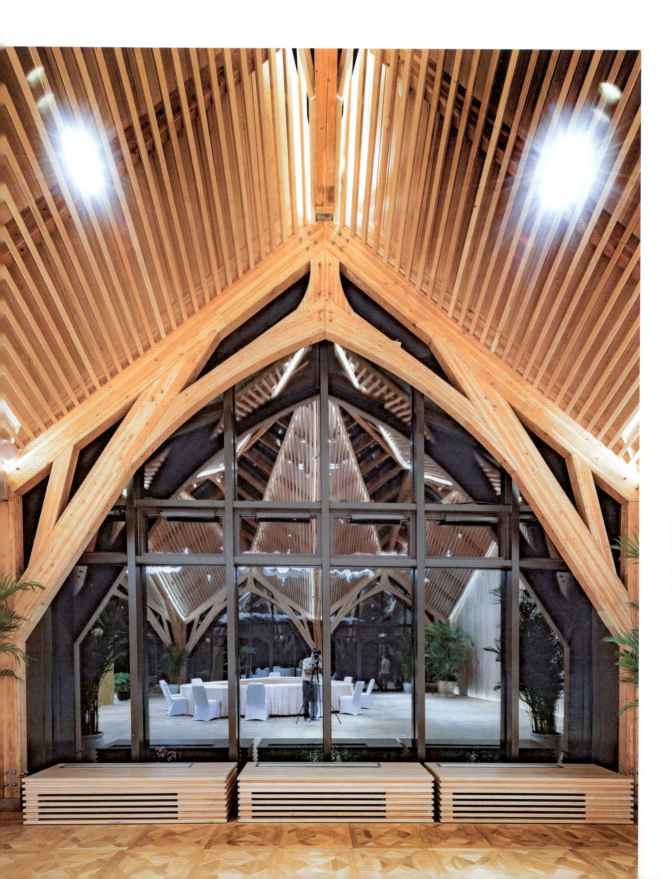

建筑以较小的建造代价，塑造了精彩的空间与迷人的形式，充分体现了木结构自然、温暖的质感，以及几何学与力学的美感。

建筑室内装饰也以木材为主要材料，通过木格栅疏密相间的设置，呈现出中国古代写意山水画的线条与肌理，隐喻了中国传统中对宁静、悠远的自然环境的向往，及其所代表的"天人合一"的理念与暗含的智慧与品德。

The entire building uses about 212 cubic meters of wood, all of which are imported timber from Canada, with a total potential carbon reduction of about 557 metric tons. Among them, roofs and infill walls use SPF (No.2) about 57.81 cubic meters, increasing the overall building insulation and energy efficiency performance. The main structure frame uses 108.57 cubic meters of Douglas Fir, the structural grade, providing high strength and beautiful texture, which increases the aesthetics of the building space while ensuring the safety of the structure. The ceiling and wall decorations use 42.68 cubic meters of Hemlock with fine texture.

Design of this building began in December 2016 and construction started in March 2018 and finished in October 2018. Throughout the design and construction process, the structural engineers from Canada Wood provided active and effective assistance in cost consulting, structural calculation, detail design, on-site quality control, etc., to get the successful completion of the building.

整幢建筑木材用量约212立方米，其中SPF 57.81立方米，花旗松108.57立方米，铁杉42.68立方米，潜在碳减排总量约557公吨。

建筑于2016年12月开始设计，2018年3月奠基动工，10月初全部建造完成，总历时23个月。在设计与建造全程，加拿大木业协会的结构工程师提供了造价咨询、结构计算、节点设计、现场指导等积极且有效的协助，共同促进建筑的成功落成。

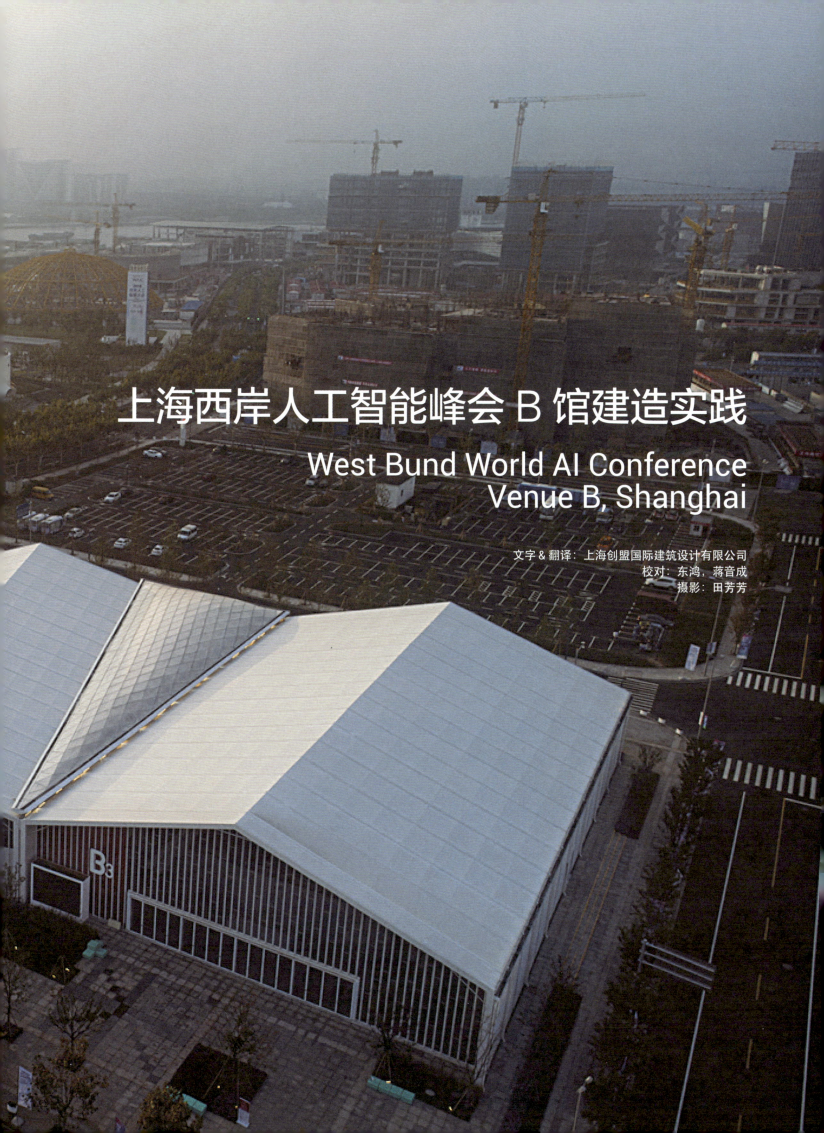

上海西岸人工智能峰会 B 馆建造实践

West Bund World AI Conference Venue B, Shanghai

文字 & 翻译：上海创盟国际建筑设计有限公司
校对：东鸿，蒋音成
摄影：田芳芳

Project: West Bund World AI Conference Venue B
Location: Longtengdadao Road, Xuhui District, Shanghai China
Developer: Shanghai West Bund Development Group Co., Ltd.
Design: Archi-Union Architecture Design Co., Ltd.
Contractor: Hongrun Construction Group Co., Ltd.
Digital Fabrication: Fab-Union Architectural Technology and Digital Fabrication Co., Ltd.
Landscape: Shanghai EcoG Garden Co., Ltd.
Floor Area: 8885m²
Design Stage: 2018.04 - 2018.07
Construction Stage: 2018.05 – 2018.09
Principal Architect: Philip F. Yuan
Design Team:
Architecture: Li Han, Jinxi Jin, Lei Lin, Jinyu Huang, Xiao Zhang
3D Robotic Printed Pavilion Design: Ce Li, Chun Xu, Sijie Gao, Zhenxiang Huang
Interior: Fuzi He, Jingyan Tang
Structural Engineer: Zhun Zhang, Junchao Shen, Tao Huang, Rui Wang
MEP: Ying Yu, Yong Wang, Dawei Wei
Digital Fabrication: Wen Zhang, Xuwei Wang, Yong Peng, Liming Zhang, Ce Li, Zhimin Wan, Chun Xu, Sijie Gao, Zhenxiang Huang

Shanghai West Bund Construction, the leading development for future urban architecture, has taken on a new challenge this year - the World Artificial Intelligence Conference (WAIC) will be held here on September 17-19, 2018. As one of the main venues for the 2018 World AI Conference, the design for Venue B was initiated in April 2018, followed by the construction in June. Finally the whole 8885 m² space was successfully realized in September.

Through algorithmic technologies such as the Internet, virtual reality, and robotic intelligence, AI has been rapidly refreshing our perceptions about the world. How can we present an urban public exhibition space, which carries the essence of AI technology in 100 days? When AI era has arrived, can this design process respond to the conceptual and technological potentials of our future construction industry?

The Building is located next to the vibrant waterfront of Xuhui Riverside in the heart of Shanghai. It is adjacent to the West Bund Art Center (Building A) and many other art spaces such as the West Bund Art Museum, the Tank Shanghai Art Parkand the Longhua Heliport. Along with the gathering of artificial intelligence conference and technology industries, the integration of art and technology has become a new topic of riverside urbanizations at the West Bund. Through simple form, pragmatic construction, cyborg ingenuity (man-machine cooperation) and fully prefabricated structure system, we are able to quickly realize the green, industrialized and intelligent architecture, and fully present the systematic solution for digital construction.

地点：上海市徐汇区龙腾大道
建设单位：上海西岸开发（集团）有限公司
建筑设计：上海创盟国际建筑设计有限公司
施工总承包：宏润建设集团股份有限公司
数字建造：上海一造建筑智能工程有限公司
景观：上海溢柯园艺有限公司
建筑面积：8885 平方米
设计时间：2018 年 4 月— 2018 年 7 月
建造时间：2018 年 5 月— 2018 年 9 月

设计团队：
主创建筑师：袁烽
建筑：韩力，金晋磎，林磊，黄金玉，张啸
咖啡亭设计：李策，徐纯，高思捷，黄桢翔
室内：何福孜，唐静燕
结构：张准，沈俊超，黄涛，王瑞
机电：俞瑛，王勇，魏大卫
数字建造：张雯，王徐炜，彭勇，张立名，李策，万智敏，徐纯，高思捷，黄桢翔

引领未来城市建筑发展的西岸滨江建设今年迎来了一项新的挑战——世界人工智能大会（WAIC）于 2018 年 9 月 17—19 日在上海西岸举行。作为 2018 世界人工智能大会的主会场之一，西岸峰会 B 馆设计工作从 2018 年 4 月启动，建造实施也紧凑地于 2018 年 6 月展开，并在 2018 年 9 月顺利实现了 8885 平方米的空间呈现。

在人工智能通过互联网、虚拟现实、机器智能等算法技术飞速刷新我们认知的时候，如何在 100 天里呈现一个可以承载人工智能技术集中展示的城市公共展览空间呢？这个呈现过程是否可以在人工智能已然到来的时代，回应我们建筑产业将如何实现观念与技术的改变呢？

西岸峰会 B 馆位于上海中心城区徐汇滨江非常具有公共活力的滨水带旁，紧邻西岸艺术中心（A 馆），与西岸美术馆、油罐艺术公园、龙华直升机场等西岸艺术空间为邻。伴随着人工智能大会与科技产业的集聚，艺术与科技的融合成为西岸滨江城市建设的全新话题。我们通过简练的形式，务实的建造，人机协作的巧思和全预制装配结构体系，快速实现了建筑工业化、绿色化以及智能化的具体实践，完整呈现了建筑绿色智能建造的系统化解决方案。

INTERVAL: Shared Gardens In Between Buildings

The twist of the three main buildings of Venue B creates two triangular park entrances – a shared urban green space with shelter and semi-openings. After stepping into this garden, you can immediately feel the warmth of the timber structure. The blend of white and wood is an unexpected surprise, and it is also a space that makes people relaxed and willing to stay and interact.

After considering functional positioning, usage analysis, maintenance, security management and different usages in the future for the three main functional spaces, the two public parks express two major scenarios. When they are fully opened, they become a pocket park within the city fabric and when they are semi-opened, they indicates the circulation and stitches together the three main volumes. While they serve as rest stations in between the summits, they are also buffering the internal traffic.

In addition, we designed a 120 m² translucent 3D printed coffee pavilion within the larger garden and nearly 50 seats to form a space for rest, communication and tea breaks during the summit. While enhancing the functionality of the space, the pavilion also adds in a new level of spatial personality and feature.

间隙：建筑中的共享公园

西岸峰会 B 馆的三个主体量的扭转形成了两个三角形的绿色入口公园——有遮蔽、半开放的共享城市空间。进入这个空间，木结构的网壳屋顶带来扑面而来的温暖气息。白色与木色的交融既是一种意外惊喜，也是让人欣喜并愿意停留、交往的空间场景。

经过功能定位与使用状况分析，综合考虑维护与内部安全管理的需要，未来三个主要功能空间会出现不同使用场景，一个、两个或三个空间都可能出现单独使用的状况。这两个共享公园会出现两种不同的应用场景：全开放时，将门全部打开形成城市口袋公园；半开放时，则变成组织内部流线的入口空间与内部公共空间，将三个分散的体量在城市空间和界面上连成一体，同时为大规模会务的人流量和辅助功能提供缓冲和体憩空间。

另外，在其中相对较大的共享花园中，我们设计了一个 120 平方米的半透明 3D 打印咖啡厅以及近 50 个座椅区域，形成峰会席间休息、交流以及茶歇等功能的空间，同时提升空间场所的全新个性与场所特色。

COVERAGE: Warm Parametric Timber Structure

The shared garden is separated into two triangles, which are naturally defined as one moving and one static space. Both courtyards are covered with digital prefabricated timber vaulted roof, in which the larger one has a span of 40 meters and its structural thickness is only about 0.5 meters. It is the most economical inter-supported steel-wood roof in the world.

Through algorithms, the form is slightly arched and the force is evenly distributed. The overall arch is balanced by the lateral arrangement of the steel trusses and is further reinforced at the three corners. The inner timber arches are optimized to the double-hollow superimposed beams, and the geometrical dimensions of all the beams are further refined by digital form-finding, so that every single beam can be optimally materialized during gluing. The installation process only requires 3~4 labors, which has significantly improved the construction efficiency on site. All beam heads are optimized in a parametric manner and such data is then used to guide the digital fabrication for milling and boring. The joints feature a standardized hollow aluminum structure that further reduces roof weight while facilitating prefabrication and on-site construction.

On-site construction for the 2000 square meter timber shell only took 29 days. From the side, the ceiling is slightly above the main conference space, which results a better ventilation on ground level. The top is covered by polycarbonate corrugated board. Light is filtered through several layers and sprinkled in the gardens.

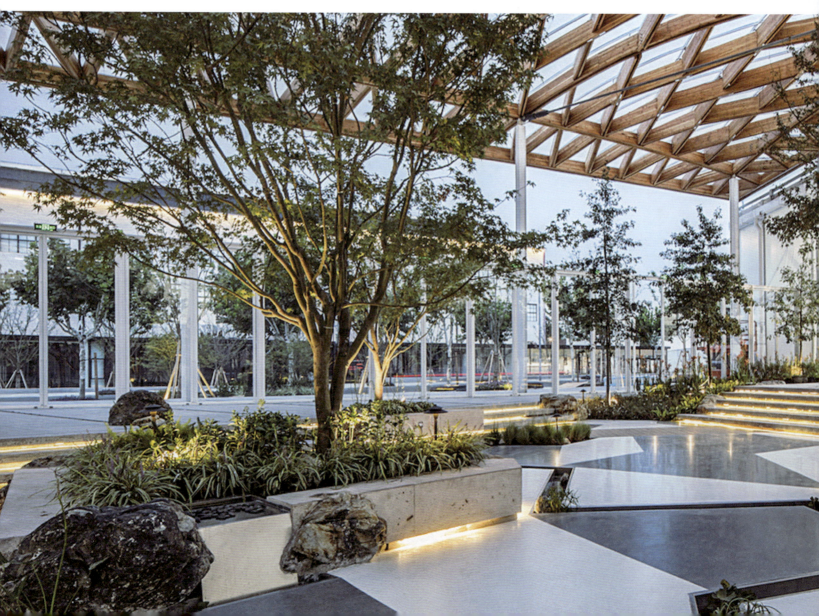

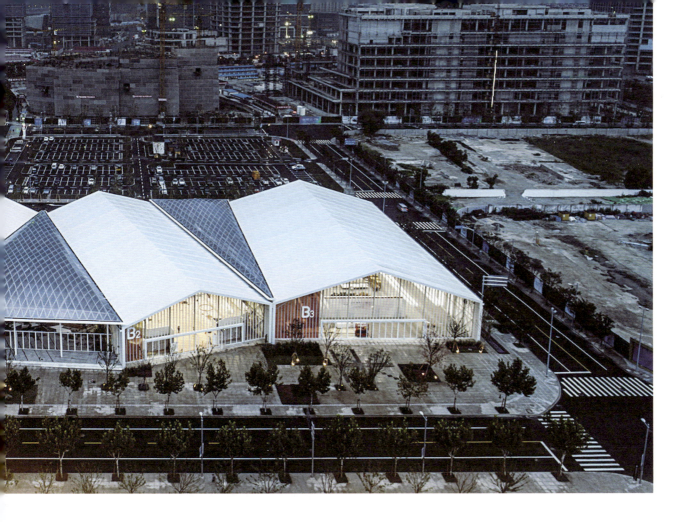

覆盖：温暖的互承式参数化木构

共享花园被主体量自然地围合成一大一小两个三角形，从空间特性上也自然地界定为一动一静，都通过数字预制化木构拱壳顶棚加以覆盖；大庭院屋顶跨度约为40米，结构厚度仅为0.5米，是全球单元材料最省的互承式钢木结构屋顶。

通过数字化的力学找形，木构屋顶整体微微起拱，实现了力学的合理布置。整体木构拱架的侧推力通过周围的横向布置钢桁架得以平衡，并在三个角部通过局部加强。内侧木构拱壳全部优化为互承方式的双幅中空叠合式木梁，并通过计算机二次找形优化所有木梁的几何尺寸，使得单根木梁在胶合时做到材料最优，安装时3~4个工人就可以搬运，也提高了现场的安装效率。所有的梁头通过参数化的方式进行非标设计优化，并通过平行数据指导数字工厂进行节点数字化铣削和开孔加工。连接件采用了标准化的中空铝构造，进一步减轻了屋顶重量，同时方便预制加工与现场施工组织。

2000平方米的木壳部分现场仅29天的施工周期，全部通过预制拼装的方式得以实现。顶棚在侧面微微高于主会议空间，这样可以形成更好的地面通风效果。顶部通过聚碳酸酯瓦楞板加以覆盖，光线经过几层过滤后洒在共享花园中。

INTELLIGENCE:
Fully Prefabricated Modular System

West Bund World AI Conference Venue B attempts to redefine the entire process from design to construction. The data model has replaced the traditional format of drawings and become the medium for form, structure, prefabrication and on-site installation. Through digital form-finding, parametric optimization and parallel date for fabrication and construction, we are trying to redefine different ways of intelligentization in all aspects of architecture, and realizing the integration of intelligent design and construction.

The main space uses prefabricated light aluminum truss system, which is a mature system and allows for accurate construction. It is also the building system with the lightest unit weight from all the known materials. Considering the need of rapid construction, the building model and façades treatments are trying to employ conventional products. At the same time, in order to improve the urban integrity of the main façade, a refined cross-steel keel system and a semi-concealed polycarbonate curtain wall are added into the design. The keel treatment creates a translucent spatial texture, elegantly transforming and filtering the urban space into the exhibition space. The main entrance is slightly retracted, and the extension of the roof is used to form a gateway for the entrance gallery, indicating the openings of the main façade. The scale however, does not show an encouragement for people to stay too long at the entrance and thus is able to manage the traffic along the street.

The interfaces on each side are modified according to the needs of operation. For instance, the facade to the main street is closed as much as possible to avoid noise and interruptions to the summits. The façade connecting to gardens however, is widely opened up to allow relaxation within light and nature in between conferences. A mature multi-layer composite roofing system is used to enhance the energy performance of the entire project while meeting the acoustic characteristics.

智造：模数化结构几何的全预制化建造

西岸峰会B馆尝试重新定义从设计到建造的整个流程；数据模型在一定程度上取代了通常意义上的图纸成为形式、结构、预制加工和现场安装的媒介；通过数字化智能几何找形、参数化力学建造优化方法以及平行数据指导数字工厂加工和建造的方法，尝试重新定义建筑各个环节智能化的不同推进方式，实现智能化设计建造一体化实践。

主空间采用了轻铝排架预制体系，轻铝结构具有体系成熟、施工精准、配套成熟的特点，而且是目前已知的单方材料重量最小的遮蔽建筑体系；考虑峰会快速搭建的需要，建筑模数和典型立面尽量采用了常规化的体系内产品；同时为提高主立面的城市完整度，通过精致化的十字钢龙骨体系和聚碳酸酯板结合的半隐框幕墙来提升立面品质；十字化的龙骨处理形成了半透明的空间质感，优雅地实现了城市空间和会展空间的切换和过滤；主入口略微收进，利用屋顶空间的延伸形成入口廊下空间，暗示主立面入口所在，但超人的尺度又不鼓励人做过多停留，以配合大量人流进出的管理需要。

主空间侧面的空间界面考虑运营的需要选择不同的封闭属性，临向城市街道的界面尽量封闭，隔绝噪声以及城市环境对会务活动的干扰；临向共享花园的界面尽量开敞，使得紧张的会务之余，人们可以从感知自然变化，光影婆娑。屋顶选择了成熟的多层复合的屋顶覆膜系统，这样在满足会务需求的声学特性同时，提升整个项目在能耗方面的标准。

2-2 剖面图 2-2 Elevation

立面图 Elevation

立面图 Elevation

天津欢乐谷演艺中心

Art & Performance Center of Happy Valley, Tianjin

作者：贺雄岩，Shrey
翻译：张云川，Susan
校对：东鸿，蒋音成
摄影：贺雄岩，Shrey，张云川，Susan

Architect: ROMERO PETRILLI VANRELL & ASSOCIATES
Timber Structure Engineering: Shanghai Zhenyuan Timber Structures Engineering Co., Ltd.
Location: TianJin, China
Gross Floor Area: 6100m²
Number of Floors: 1
Completion Date: 2013.6

项目设计：ROMERO PETRILLI VANRELL & ASSOCIATES

木结构设计：上海臻源木结构设计工程有限公司

项目地点：中国天津

建筑面积：6100 平方米

层数：单层

竣工时间：2013 年 6 月

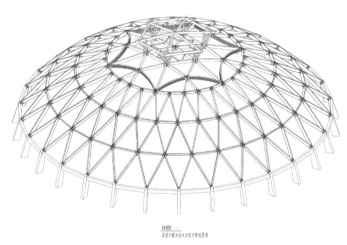

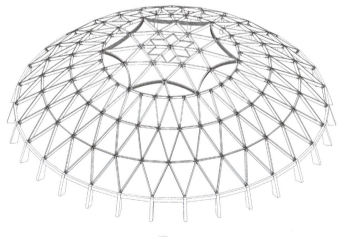

This single-layer spherical reticulated shell structure is located in a theme park. The external framework is designed to apply the Southern Pine Glulam structure with a vault. The spherical radius is 57m. The height of the chord is 19m. The bottom diameter of the wooden dome is 85m. It is supported by 6-meter-high concrete columns (and concrete ring beam).

In order to ensure the overall stability of the structure, the base joint is designed to be hinge joint, and the other joints use fixed joints. A total of approximately 360 m³ of 24F-V5 American Douglas fir Glulam is used in the building.

这座单层球面网壳结构的建筑位于一个主题公园内。网壳框架采用南方松胶合木按球面曲线设计，球体半径57米，弦高19米，木穹顶底边缘圆直径达85米，下部有6米高混凝土柱及圈梁提供支撑。

设计中为了保证结构整体的稳定性，支座节点设计为铰接，其他节点为刚接。该建筑共计使用了约360立方米的24F-V5美国花旗松胶合木。

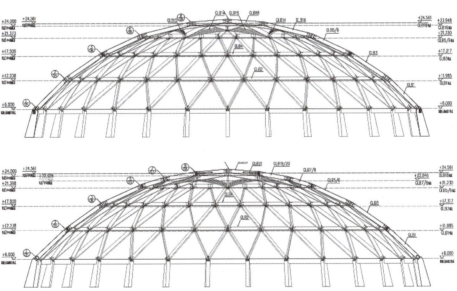

According to a certain rule, a large number of different geometric figures have been formed by the glulam members, forming a roof structure like a huge petal. The huge space inside the shell can be used as a usual performance place, or it can be flexibly divided and arranged as needed for holding exhibitions and other purposes.

The outer edge of the roof is a tempered glass roof. The transparent glass roof and the metal roof are alternately connected to provide lighting and increase the layered effect of the roof. There is a roof skylight system that can be closed in the middle of the roof. It can be opened in days of good weather to make the building fresh, reducing energy consumption and providing partial light.

室内胶合木按一定规律形成了大量不同的几何图形，形成了一个犹如巨大的花瓣的屋面结构。壳体内部巨大的空间可作为平时的演出场所，也可以灵活分割，按需要布置，以供举办展览等用途。

屋面外沿部分为钢化玻璃屋面，透明的玻璃屋面与金属屋面交错衔接，提供采光的同时增加了屋面的层次效果。屋顶中心为可闭合的电动天窗系统，在天气良好时可以打开通风，减少能耗，同时提供部分采光。

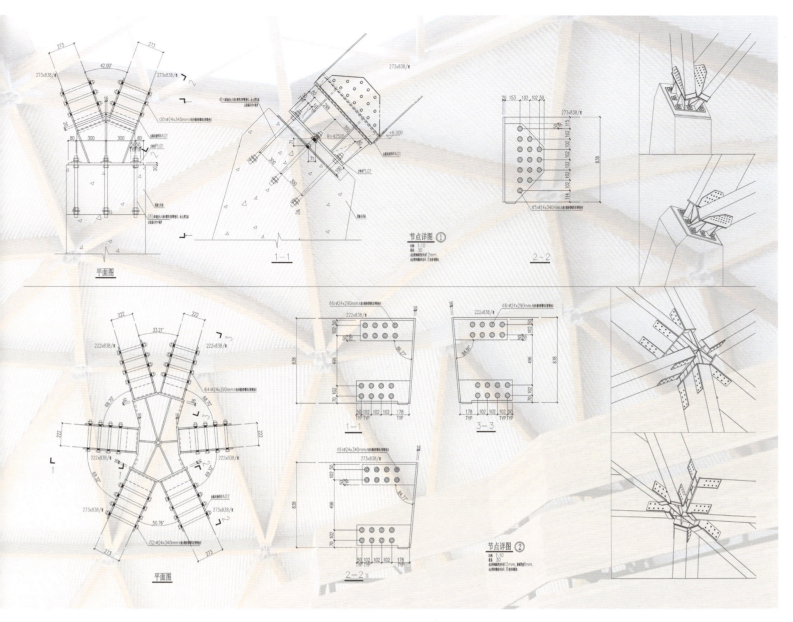

Architect: Beijing YIKE Architecture Co., Ltd.
Timber Structure Engineering: Shanghai Zhenyuan Timber Structures Engineering Co., Ltd.
Contractor: Shanghai Zhenyuan Timber Structures Engineering Co., Ltd.
Location: Ziyun Autonomous County, Guizhou Province, China
Gross Floor Area: 740 m²
Number of Floors: 1
Completion Date: 2019.7

项目设计：北京一棵建筑设计有限公司
木结构设计：上海臻源木结构设计工程有限公司
木结构施工：上海臻源木结构设计有限公司
项目地点：贵州省紫云自治县
建筑面积：740 平方米
层数：单层
竣工时间：2019 年 7 月

贵州紫云自治县格凸河攀岩基地观赛广场

Spectator Square of the Climbing base, Getu River, Ziyun Autonomous County, Guizhou

作者：贺雄岩，Shrey
翻译：张云川，Susan
校对：东鸿，蒋音成
摄影：贺雄岩，Shrey，张云川，Susan

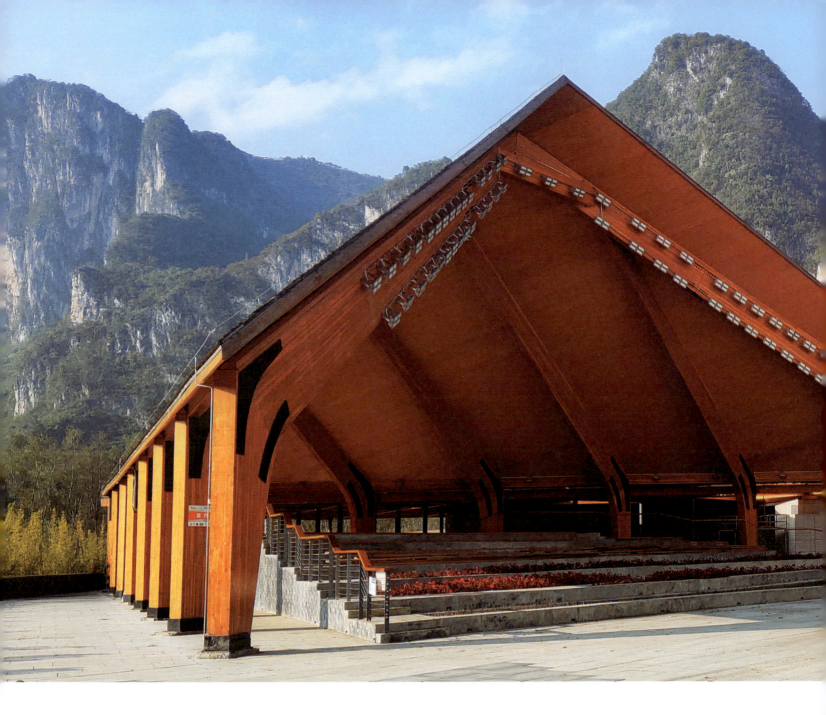

The 40-meter-long triangular pyramid structure forms the shape of the building as a semi-open pyramid. The overall structure of the building is composed of 8 groups of prefabricated arch beams with a maximum section of 1700mm and a gradual height, which maximizes the visual space.

140mm NLT structural panels laid on the roof provide heat preservation and insulation. therefore, there is no need for the ceiling decoration. The structural deformation is fully taken into account in the design and manufacturing process to ensure that all nodes are closely connected, and the butt joint of the three-dimensional arch structure is completed within the error of millimeters.

The maximum span of the building is 40m, and the height of the building top is 17.23 m. A total of 110 m³ of American Douglas fir-Larch glulam are used.

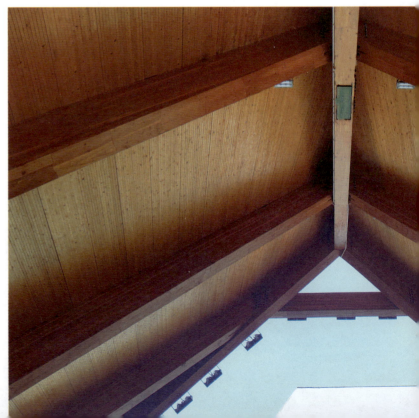

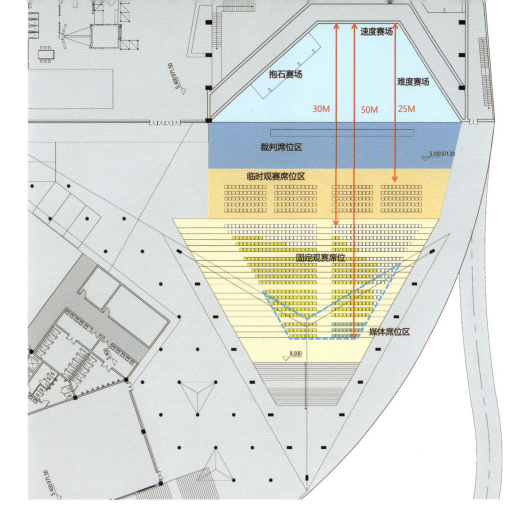

建筑外观由底边长 40 米的三棱锥结构形成建筑半开敞金字塔造形型。建筑整体结构由 8 组最大截面 1700 毫米高渐变预制拱梁对顶形成，最大限度地保留了视觉空间。

屋面铺设 140 毫米厚 NLT 结构板，起到屋面保温隔热作用，同时屋面不再需要吊顶装修。设计及加工制造过程中充分考虑到结构形变，确保所有的节点紧密结合在一起，以毫米精度完成三维拱门结构对接。

建筑最大跨度 40 米，建筑顶高度 17.23 米。共使用美国花旗松胶合木 110 立方米。

The building of this project has a beautiful shape. The color of the timber is perfectly integrated with the surrounding environment. As auxiliary facilities of the rock climbing base, the Spectator Square and the Visitor Center are closely connected through the corridor.

The building is open on all sides except the roof. The building has a good view towards the surrounding landscape. You can see the local scenic spot Guanyin Rock a little to the left. The building facing the artificial climbing wall is used as the Spectator Square during the competition days and serves as a tourist recreation square during the rest of the time.

　　本项目建筑造型优美，木材自身的色调与周边环境完美融合。作为攀岩基地配套项目，观赛广场与游客中心通过连廊紧密连接。

　　建筑仅仅屋面做了封闭，四面开敞，身处观赛广场内，视野良好，周边基地景色一览无余，稍稍左望就能看到当地的特色景点观音岩。建筑正对面是人工攀岩壁，在举行比赛期间，作为观赛广场；在非比赛时间，建筑作为游客休闲娱乐广场。

江苏省委会议厅

A Conference Hall in Jiangsu Province

作者：刘杰、曹晨
翻译：沈姗姗
校对：东鸿
摄影：叶虎

In recent years, the government has issued a number of policies on the development of timber structure architecture. In 2017, the Jiangsu Provincial Government also proposed the policy of "while vigorously developing prefabricated concrete buildings, it needs to promote prefabricated steel structure and prefabricated timber structure architecture" (Provincial Government's Opinions on Promoting the Reform of the Construction Industry, 2017).

In this context, the project follows the theme of the development of the contemporary construction industry. It combines the characteristics of the project with local conditions. It is determined basic design principle that the traditional timber structure is adopted for the landscape architecture, and the modern timber structure is used in the conference hall. It creates harmony between the ancient and modern architecture, and also between the architecture and landscape.

1 Project Overview

This project is located in an institution community in Nanjing, Jiangsu Province, and is one of the new construction projects in the Beishan Landscape Renovation Project. Beishan Mountain is a small soil slope in this community. It is high in the southern side and low in the northern side, surrounded by roads. The height difference between the highest point of the soil slope and the roads on the south and north is about 5m and 10m respectively. There also has existing buildings in the north, northeast and southwest corners.

The theme of the entire Beishan Landscape Renovation Project is based on Yangzhou's traditional garden style. New walkway, folding corridors, and Simianbafang Ting (the pavilion, with four-faced and eight-sided) with local characteristics of Yangzhou are built on the slopes. The mountains are piled with stones, and paths, buildings and landscape are built together. These form the landscape space full of characteristics.

The southwest corner of Beishan Mountain is the proposed site for the conference hall. The two-story office building with brick-concrete structure was planned to be demolished to construct a new conference hall on the original site. At the beginning of the design, the owner wanted to build the new conference hall, which could make full use of the surrounding landscape and environmental conditions, with certain characteristics.

The completed conference hall is single-story, with a rectangle-shape plan, 20.10 m in east-west length, 12.20 m in north-south width, and a building area of about 242.00 m^2. It has an overhanging gable roof and west gable wall with cornice. The height of the building is about 6.68 m and the total height from the ridge to ground is about 9.60 meters. The building is designed as a small conference hall, can hold 110 people with service rooms such as control rooms, lounges, and bathrooms.

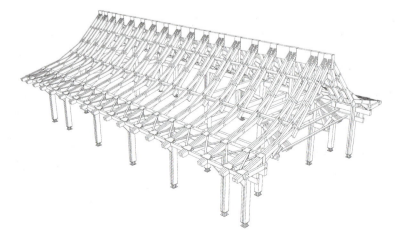

近年以来,国家先后出台若干有关发展木结构建筑的政策。2017 年江苏省政府也提出"在大力发展装配式混凝土建筑的同时,积极推广装配式钢结构建筑和装配式木结构建筑"(《省政府关于促进建筑业改革发展的意见》,2017 年)。

在此背景下,本项目紧扣新时代建筑业发展的主题,并结合项目特点,因地制宜、汲古扬今,确定了本项目中景观园林建筑部分采用传统木结构、会议厅部分采用现代木结构的基本思路,古今相映,意图做到建筑与景观的相得益彰。

1 建筑概况

本项目位于江苏省南京市,是作为北山景观提升项目中的新建工程之一。北山是院内一座规模不大的小土坡,南高北低,四面有道路环绕,土坡最高处距南、北侧道路高差分别约 5 米、10 米,北侧、东北角、西南角均有建筑物。

整个北山景观提升项目以扬州古典园林风格为主题,在山坡上新建游步道、折山廊以及富有扬州地方特色的"四面八方亭"等,并掇山叠石、辟径筑路,营造建筑、山水之景,形成富有特色的景观园林空间。

北山西南角是会议厅的拟建地,原有一座两层砖混结构的办公楼,规划拆除后在原址上建设成新会议厅。设计之初,业主单位也希望新会议厅能够充分利用周边的景观环境条件,并具备一定特色。

建成后的会议厅为单层,建筑平面呈矩形,东西长 20.10 米,南北宽 12.20 米,建筑面积约 242.00 平方米,悬山顶,西侧山面带披檐,建筑高度约 6.68 米,屋脊至室外地面高度约 9.60 米。本建筑定位于小型会议厅,设定人数为 110 人,配套控制室、休息室、卫生间等服务用房。

2 Design Concept

The architectural design concept of this conference hall continues the thought of Beishan Landscape Renovation Project, combined with the functional requirements of conference activities, and adapts to the entire Beishan landscape and its cultural environment. It is intended to build a modern building both with Chinese traditional architectural style and spirit of times. This design draws design inspiration from traditional Chinese construction thoughts and wisdom, with concise and transformation. It uses modern architectural design language to express the architecture in the new era.

One of the characteristics of the traditional Chinese construction thought is "small materials constructed for large span structure", which means that small-sized components are combined to form a stable structural system. For example, Dougong built in traditional official buildings is made by the combination of a series of standard components, and is used as important supporting and connecting components for the building.

Another example is the Rainbow Bridge in the *Changming Shanghe Tu*, which is a long-span structure bridge constructed by many small section logs stacking vertically and horizontally. Its span is about 20 meters long, which meets the navigation requirements. The structure of the Rainbow Bridge was not recognized by the Chinese history of science and technology until the 1980s, and gradually revealed the scientific nature of the structure of the Rainbow Bridge, and named it *guanmugong* (Intersecting timber arch) or *bianmugong* (woven arch) structure, named by Author. Then, he successively explored the timber arch covered bridges which are common built in Zhejiang and Fujian Provinces by the structure similar as the Rainbow Bridge.

In brief, the structure of the Rainbow Bridge is composed of two sets of systems that are alternately combined laterally side by side. System 1 is composed of three long rods and system 2 is composed of two long rods and two short rods. The rods are connected with the horizontal timber by mortise and tenon joint, and components from the two systems are interspersed with each other to form a stable structural system.

From the structural characteristics of the Rainbow Bridge, it is very similar to the folk weaving technology, so the author named the Rainbow Bridge's structure as a structure of Woven Arch. From this point of view, the Rainbow Bridge's structure is naturally wearable, and we can even make this Woven Arch into a ring to form a ring structure or a space structure.

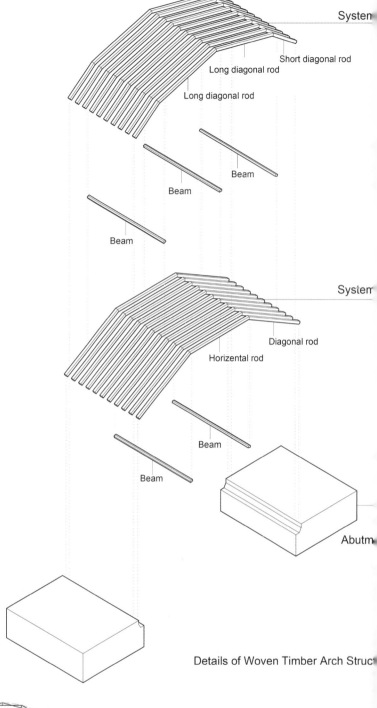

Details of Woven Timber Arch Structure

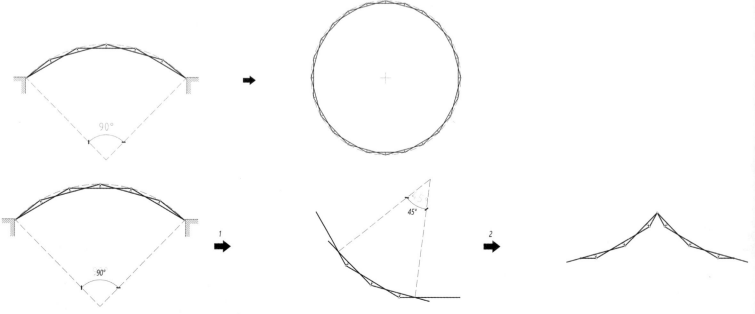

2 设计构思

本会议厅建筑设计构思延续北山景观营造的思路，结合会议活动的功能需求，适应整个北山的景观、人文环境，意在建设一座既具中国传统建筑风貌又具有时代精神的现代建筑。本设计从中国传统营造思想与智慧中汲取设计灵感，并加以提炼、转译，用现代的建筑设计语言来表达新时代的建筑。

中国传统营造思想的特征之一是"小材大用"，即以小尺寸的构件组合成稳定的结构体系。如传统官式建筑常见的斗栱，就是通过一系列标准构件组合而成的组件，而作为建筑中重要支承与连接构件。

又如北宋名画《清明上河图》中的"汴水虹桥"，就是运用小断面的圆木通过构件之间纵横叠压，而形成长达近 20 米的大跨结构，并满足了古代漕船的通航要求。"虹桥"结构直到 20 世纪 80 年代，才被中国科技史学界所认知，并逐步地揭示了"虹桥"结构的科学性，并将其命名为"贯木拱"或"编木拱"（笔者）结构，之后又陆续探寻出疑似"虹桥"结构遗存的浙闽地区常见的木拱廊桥。

简单来看，"虹桥"结构是由两套系统横向并排交替组合而成，系统 1 是由三长杆构成，系统 2 是由两长杆、两短杆构成，各杆件之间通过榫卯与横向的木方连接，两系统构件之间相互穿插别压，形成一个稳定的结构系统。

从"虹桥"结构的构造特征来看，非常类似于民间的编织工艺，所以笔者将"虹桥"结构命名为"编木拱"结构。从这个角度来看，"虹桥"结构具有天然的可编织属性，所以我们甚至可以将此"编木成环"，形成一个环形结构或者空间结构。

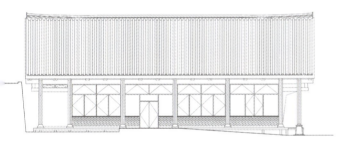

平面图 Plan

立面图 Elevation

On the other hand, a major feature of traditional Chinese architecture is the sloping roof, which has many different types. Regardless of a civil building or an official building, it is basically in the form of a pitched roof with a soft concave curve. The load-bearing structure of roof purlin-rafter is the internal reason for its roof lifting.

Therefore, this design combines the characteristics of both, and innovatively designed the roof of the conference hall. Through the decomposition, recombination, inversion, and combination of the structure of traditional timber arch bridge, the convex arch curve is creatively turned into the concave curve which is the traditional roof curve. The design method shows a unique structural system and design concept, and uses a small cross-section engineering wood and metal connections to achieve the perfect integration between the structure and the shape. And it forms a large-span column-free space.

The function of the entire conference hall is relatively simple. It is simply divided into three parts: the front hall, the main hall, and the backstage. The left and right sides of the front hall are the rest room, the bathroom and tea room, and the hallway is in the middle; the main hall can fit 110-seater and a rostrum for 7 people. The backstage is located on both sides of the rostrum, which is the control room of the strong and weak electricity room of the entire conference hall. Large glass windows are set on the north and south sides of the building and it can easily enjoy the outdoor landscape from the south side.

In the construction of traditional Chinese timber architecture, there was no professional subdivision between architecture and structure. Craftsmen (builders) are masters who integrate design and construction skill. The shape, scale, structure, and decoration of the building are all controlled by the craftsmen. In some certain sense, the beauty of traditional Chinese architecture lies in the harmony of architecture and its own structure system. It is difficult to distinguish between architecture and structure, or architecture as structure. Therefore, in this design, we have also inherited the traditional construction method of *qishangmingzhu*, which makes the roof truss directly exposed, showing the beauty of the structure and the material.

In this design, in order to highlight the continuity of the appearance of the timber frame, all steel connection nodes are concealed except for the exposed bolt heads decorated with stainless steel nuts.

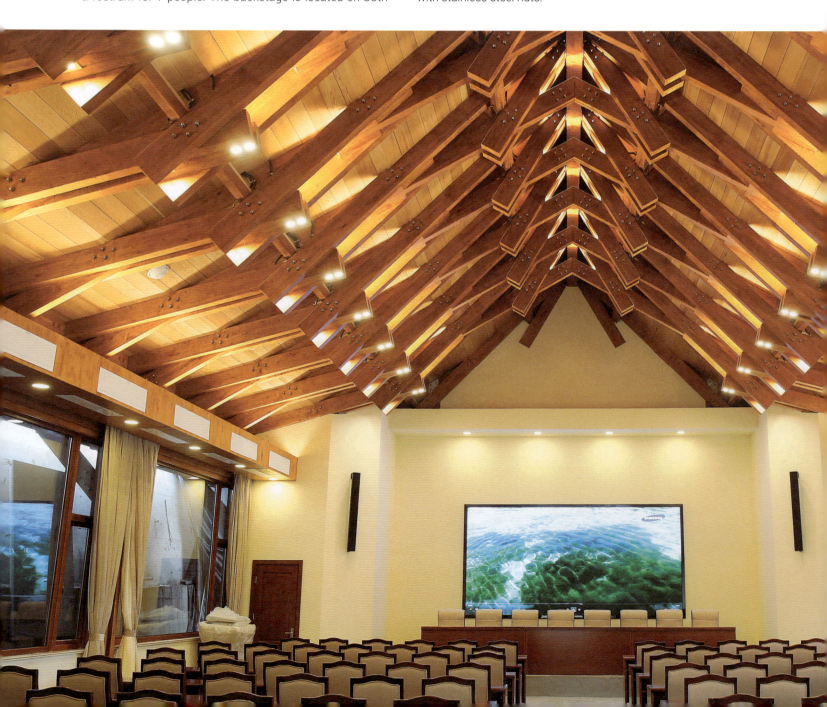

另一方面，中国传统建筑一大特征就是坡屋顶，且类型众多。不管是民间建筑，还是官式建筑，基本都是坡屋顶的形式，且带有柔和的内凹曲线，屋顶檩-椽承重结构形式是屋面举折生成的内在原因。

因此，本项设计结合两者特点，对会议厅的屋顶进行了创新设计，通过对传统木拱桥结构的分解、重组、反转、组合的方式，创造性地将外凸的拱曲线翻转成内凹的传统屋面反曲线，展现出独特的结构体系与设计理念，并应用小断面的工程木材与金属件连接的方式，从而实现结构与造型的完美统一，实现了大跨无柱的空间。

整个会议厅功能空间上比较简单，简单划分为前厅、正厅、后台三个部分：前厅左右两边分别是休息室与卫生间、茶水间，中间是过厅；正厅是110人的座席与7人的讲席；后台位于讲席两侧，是整个会议厅的强弱电间控制室。建筑南北两侧均设置大玻璃窗，南侧可方便地观赏到室外的景观绿化。

中国传统木结构建筑营造中，没有建筑、结构等专业细分，工匠是集设计、建造于一身的集大成者，关于建筑的造型、尺度、构造、装饰等方面均由工匠把握。某种意义上看，中国传统建筑之美在于建筑与结构的和谐之美，我们很难区分哪里是建筑、哪里是结构，抑或讲建筑即结构，结构即建筑。由此在本项设计中，我们也是继承了"彻上（露）明造"的营造做法，直接露明屋架，展现结构之美与木材之美。

本项设计中，为突出木构架的外观连续性，除必要的裸露螺栓头用不锈钢螺帽装饰外，所有钢连接节点均进行隐蔽设计。

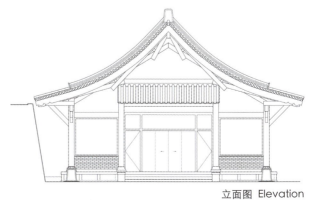

立面图 Elevation

3 Structural Design

The structural materials of the entire conference hall are made of high-quality Douglas fir glulam. The size of the components is limited as much as possible in the design phase. The single-roof roof trusses are composed of several small sections with a section size of 100mm × 250mm. The columns are also composed by the components with cross-section dimensions of 100mm × 350mm and 100mm × 250mm, which are assembled into I-shaped cross-sections. Except for the size of some parts of beam is slightly increased, the cross-section dimension of most beams is 200mm × 350mm. It has reached the design requirements with less specification and more combinations.

The structural design of the roof truss is one of the difficulties and highlights of this design. The design inspiration was taken from the structure of the Rainbow Bridge. The original arch structure was turned into a diagonal beam structure, which seemed to be a kind of subversion. The structure of woven arch is transformed. The traditional mortise and tenon joint or binding connection between the structure components of woven wood arch was replaced by modern bolted connections, and the horizontal beam that originally interposed between the two systems were replaced by prefabricated metal connectors.

At the same time, it is considered that the two-slope roof truss structure will cause a large side thrust force to the substructure after being carried load. In order to solve this problem, the usual method is to add horizontal or cross diagonal rods between the two inclined beams. But to this project, in accordance with the design principle of highlighting the structural features and reducing visual obstruction, this design abandons the conventional methods. It achieves the same purpose by adding some special structures, such as strengthening the connection between the diagonal beam and the ridge, the diagonal beam and the column, and adding X-shaped steel supports along the cornice of roof truss. The roof trusses are connected to each other to form a truss structure to enhance the stiffness of the roof structure. Thereby the side thrust force can be reduced after the roof trusses are loaded.

3 结构设计

整个会议厅的结构用材均采用优质的花旗松胶合木，并在设计中尽可能地减少构件的规格尺寸，单榀屋架均是由截面尺寸为 100 毫米 ×250 毫米的多段小构件组合而成，柱也是由截面尺寸为 100 毫米 ×350 毫米、100 毫米 ×250 毫米的构件拼合成"工"字形截面的拼合柱，除局部梁尺寸稍增大外，多数梁的截面尺寸均为 200 毫米 ×350 毫米，达到"少规格、多组合"的设计要求。

屋架的结构设计是本项设计的难点也是亮点之一，设计灵感取自"虹桥"结构，将原先的拱结构翻转成一个斜梁结构，似是一种"颠覆"，原先分析"编木拱"结构的思路在此应予转换。"编木拱"结构构件之间传统的榫卯或绑扎连接的方式被替换成现代的螺栓连接，原本穿插于两系统之间的横木被弱化，取而代之的是预制金属连接件。

同时考虑双坡屋架结构承载后会对下部结构造成较大的侧推力，通常的方式是在两根斜梁间增设水平或者交叉斜向的拉杆，但本着突出构架特征、减少视觉遮挡的设计原则，本设计摈弃常规的处理方式，通过增设一些特殊的构造来达到相同的目的，如增强斜梁与正脊、斜梁与柱的连接，在檐口屋架上沿增设 X 形钢支架等措施，让每榀屋架彼此联系，形成类似桁架结构，来增强屋架的刚度，来以此消减屋架承载之后的侧向推力。

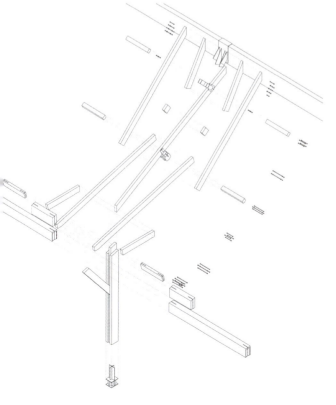

The lower structure of the conference hall adopts the frame structure with beam and column. However, if considering that the large area of windows on both sides of the conference hall will cause the insufficient lateral force resistance of the structure, the design method is to convert limited building elements into structural components as much as possible. For example, we make use of the under-window wall, which is set under the large glass window, as a timber framed shear wall to provide the side lateral force.

The entrance of the conference hall is also one of the highlights of this design. The main entrance is set on the west wall of the building, combining the characteristics of the site and the interior space of the building. It is different from traditional design method that setting the entrance of the building on its long side. In ordering to reduce the light interference and the influence of interior decoration, the cantilever of the canopy on the gable wall is 2.5 meters long, which is similar to the traditional eave structure. The form of a concave entrance is adapted to comprehensively solve the above problem.

How to realize the overhanging roof truss structure under this situation is another technical problem. As a result, the wisdom of traditional buildings inspired us, such as the structural frame of chuandou and overhanging structure commonly used in ancient buildings in the Jiangnan region. Here we carried out the transformation. The specific measure is that the variable cross-sections of the upper and lower beams pass through the columns and head out, and the metal connector at the end is fixed to the column, which forming a two-story overhang to jointly support the overhanging roof truss.

会议厅的下部结构还是采用梁、柱框架结构，但考虑会议厅两侧大面积开窗造成结构抗侧力能力不足的情况，设计中尽可能地将有限的建筑要素转换为结构构件，如将大玻璃窗下的窗下墙设置为木骨架剪力墙，提供必要的抗侧力支撑。

会议厅的入口也是本项设计的亮点之一，结合地块与建筑内部空间的特征，主入口设置在建筑西山墙面，这一点不同于传统建筑的长向入口，考虑减少太阳西晒对出入人员视线干扰及室内装修的影响，山面挑檐长达 2.5 米，并设置类似传统的披檐构造，采用凹入口的形式，全面地解决上述问题。

如何在此条件下实现屋架结构的外挑，是面临的另一技术问题？由此，传统建筑的构造智慧给了我们灵感，如江南地区古代建筑中普遍应用的"穿斗"构架形式以及悬挑构造，在此我们进行了转化设计。具体措施是上下两根梁的变截面穿柱出头，尾端金属连接件与柱固定，从而形成两层出挑，共同承托出挑的屋架。

4 Field Assemble

All the timber components of this project are manufactured and slotted in the factory. Each component is coded before leaving the factory and sticking a QR code is to facilitate factory inspection and verification. Due to the particularity of this design, the dimension of individual component is not large. For components that can meet the transportation requirements, they are assembled in the factory first. After being transported to the site, they are directly hoisted after being combined with other components, effectively improving construction efficiency. It is to ensure that the project can be completed on time and passed the project acceptance within the required time.

5 Conclusions

The design of this project is the innovative design of modern timber structure architecture, based on traditional construction thoughts. The traditional structure of the Rainbow Bridge has been interpreted, inheriting the wisdom of "small materials constructed for large span structure", and achieving a large span. It is also an architectural adaptive expression based on the natural and human environment of the project location.

The design team has been committed to the research and practice of traditional and modern timber structure architecture for more than ten years, trying to use modern

4 现场装配

本项工程的所有木构件均是在厂房内加工制作、开槽钻孔，每根构件均在出厂前进行编码，并贴示二维码，方便实行出厂检验、进场验证。由于本设计的特殊性，单根构件的尺寸均不大，对于能够满足运输要求的组件，先行在工厂里组装好，待运至现场后，与其他构件组合后，直接吊装，有效地提高施工效率，确保本工程能够在规定时间内按时竣工并通过工程验收。

5 结语

本项目的设计是基于传统营造思想基础上的现代木结构建筑创新设计，将传统的"虹桥"构造进行了传译，继承了传统建筑的"小材大用"的营造智慧，实现了大跨度的空间，同时也是基于项目所在地的自然、人文环境的建筑适应性表达。

本设计团队十余年来一直致力于传统与现代木结构建筑的研究与实践，试图用现代的技术语言来传承并发扬中国数千年以来的木构营造传统，尝试设计了如杭州香积寺、柳州开元寺等传统与现代结构的木结构建筑工程，并主动对接世界范围内木结构建筑发展前沿，期待能够呈现出更多、更好的现代木结构建筑作品，为中国的现代木结构建筑的发展之路提供思路与贡献力量。

technical language to inherit and carry forward the traditional timber construction technology in China. We has designed the projects combined with traditional and modern timber structures, such as Xiangji Temple in Hangzhou and Kaiyuan Temple in Liuzhou, and actively connected to the world's leading edge of timber structure technology. We are looking forward to showing more and better works of modern timber structure architecture, providing ideas and contributions for the development of China's modern timber architecture.

Location: Taiping Lake Forest Town, Mile, Yunnan
Function: conference, convention and banquet center
Architect: Yong Zhang, Yinian Dou, Haibo Lv, Miaomiao Tang, Jinxin Wan, Fan Liu, Guoqi Cui et al.
Structural engineer: Yinlin Zhuang, Yun Yang, Jingxi Meng, Kaikai Fei et al.
Floor area: 7500 m²

Project Brief

The International Conference Center under planning is located to the east of the main entrance of the town, facing east and the broad water view with the hills in the back. The project has been positioned as a regional conference center with high standards. The modern prefabricated large-span timber structure has been applied in the project designed jointly by Shanghai Jiao Tong University Planning and Architectural Design Ltd. and Yun Nan Jicheng Landscape Design Ltd.

The gross floor area of the conference center is 7500 m². The main building has a tripod arrangement, making the functional order closed and complete. Taiping Lake Forest Town, where the project is located, is positioned as an international eco-tourism resort and a healthy living destination. The International Conference Center embraces a broad water view in the front and lies on hills in the back. In the northeast side is a tourist service center with convenient transportation and can provide service assistance for the conference center. Near the Taiping Lake area in the south, an international five-star resort hotel and natural hot spring center is under planning.

In terms of function and technology, the conference center has a high starting point to meet the needs of modern large and medium-sized conferences, meetings and banquets, in order to build a high-quality modern conference center. The ultra-light metal tiles are used on the roof which looks beautiful. A large area of glass curtain wall and grid-like decorative wood strips are used on the flank of the roof, providing the indoor sunlight and privacy.

Taiping Lake International Conference Center in Forest Town of Mile City, Yunnan
云南弥勒太平湖森林小镇国际木屋会议中心

作者：张勇，戈冬明，汤玉辉，詹晖
翻译：蒋音成
校对：东鸿，詹晖
摄影：Allen 彭

项目地址：云南省弥勒市太平湖森林小镇
项目功能：大中型会议、会晤及宴会
建筑师：张勇，窦义年，吕海波，汤淼森，万金鑫，刘帆，崔国旗等
结构师：庄寅麟，杨芸，蒙婧玺，费凯凯等
建筑面积：7500 平方米

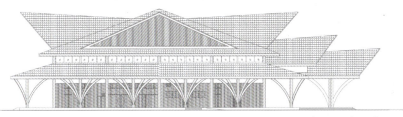

立面图 Elevation

项目简介

规划中的国际木屋会议中心位于小镇主入口东侧，背山面水，坐西向东，视野开阔。项目本着高起点规划、高标准设计的原则，定位为"区域性国际会议中心"，采用现代装配式大跨度全木结构，由上海交通大学规划建筑设计有限公司及云南吉成园林设计有限公司联合设计。

会议中心建筑总面积为 7500 余平方米，会议中心主体呈不规则"品"字形，建筑功能秩序闭合完整。项目所在地太平湖森林小镇规划定位为国际生态旅游度假区、健康生活目的地。国际会议中心背山面水，明堂开阔。东北侧为游客服务中心，交通便捷，更可为会议中心提供服务辅助。正南方临太平湖区域，规划为国际五星级度假酒店及配套天然温泉休闲中心。

在功能和技术层面上，会议中心高起点满足现代大中型会议、会晤及宴会需求，力求打造高质量现代化会议中心。建筑屋顶采用超轻金属瓦，自重轻，美观大方；侧边采用大面积的玻璃幕墙与格栅状装饰木条结合的墙体，既能保证室内光线，同时也具有隐秘性。

Theme of the project

Mile, the only city in the name of the Maitreya Buddha, has unique Buddhist cultural heritage. The lotus, which grows in the mud without being stained, is the sacred flower of Buddhism, symbolizing the transition from this world to the pure world, from evil to perfection, from ordinary to buddha. Many great civilizations such as the ancient Egypt and India take lotus as a sacred symbol and integrate it widely into their art and architecture.

Concept of the project design——creating the beauty of architecture with timber structure

The culture of wooden architecture embodies the Chinese philosophy that Tao follows the law of nature and that human being is an integral part of nature, as well as the religious complex loving wood. Wood is also the source and soul of Taiping Lake Forest Town. The town has built the largest high-end wooden house hotel group in China, with nearly a hundred prefab wooden houses.

The conference center will adopt a prefab large-span heavy timber structure, combined with a simple and modern Chinese architectural style. It will make full use of the high and transparent space, and integrate the wood culture, Buddha culture, and national culture on the premise of meeting the basic conference functions.

The Chinese simple modern architectural style is present in the design. The simplified traditional eaves, the exquisite cornice lines, the transparent floor-to-ceiling glass, the solid wood support create an eco-friendly, modern and majestic space. The interior decoration is closely linked to the theme Lotus of Peace and Maitreya Blessing. The main conference hall is based on a simple Chinese style, supplemented with Southeast Asian decorative elements, showing a unique oriental charm. The other function halls are decorated with the elements of Yunnan's national culture and interpreted ethnic customs. The overall interior presents friendship, peace, and tolerance, and fully reflects the regional, cultural, and the features of the times.

项目主题

弥勒,唯一与佛同名的城市,具备得天独厚的佛禅文化底蕴;莲花,佛教圣花,出淤泥而不染,象征着从尘世到净界,从诸恶到尽善,从凡俗到成佛,最终修成正果。从埃及到印度的众多世界伟大文明都把莲花视为神圣的象征,并将其广泛地融入他们的艺术和建筑之中。

项目设计理念——以全木结构成就建筑之美

"木结构建筑文化"体现着中华"道法自然、天人合一"的哲学观和民间"盛木为怀"的宗教情结。"木"也是太平湖森林小镇的本源与灵魂所在,小镇已建成国内最大的高端木屋酒店群,拥有近百栋装配式独立木屋。

会议中心将采用装配式大跨度重木结构,配以简约大气的现代中式建筑风格,充分利用挑高通透的空间,在满足基本会议功能的前提下,融入特色木文化、佛文化、民族文化,打造具有高识别度的高端国际会议中心。

整体方案采用中式简约的现代建筑风格,提炼简化传统屋檐,配以简约精巧的檐口线条;通透开阔的全落地玻璃,辅以大气沉稳的实木支撑,低碳环保却也现代前沿、磅礴大气。室内装饰紧扣"和平之莲、弥勒添福"主题,主会议厅以简约中式风格为主,辅以东南亚装饰元素,展示独特的东方神韵。其他功能厅则精炼云南民族文化元素,演绎民族风情。整体内装贯穿友谊、和平、包容,充分体现项目地域性、文化性和时代性。

加拿大
素里城市中心
Surrey Central City, Canada

来源：Naturally Wood
翻译：高晓萌
校对：沈姗姗
摄影：Nic Lehoux，Tae Ik Hwang

Owner: ICBC Properties
Architect: Bing Thom Architects
Structural engineer: Fast + Epp
Completion: 2003
Size: 7896 m²

Awards:
Commercial Award of Excellence, Surrey New City Design Awards
Interiors Award of Excellence, Surrey New City Design Awards
Citation Award, Wood Design Awards
Illuminating Engineering Society Lighting Design Awards
Marche International des Professionels de l'Immobilier Special Jury Prize
Lieutenant-Governor of British Columbia Medal in Architecture, AIBC
Architectural Institute of British Columbia Innovation Award

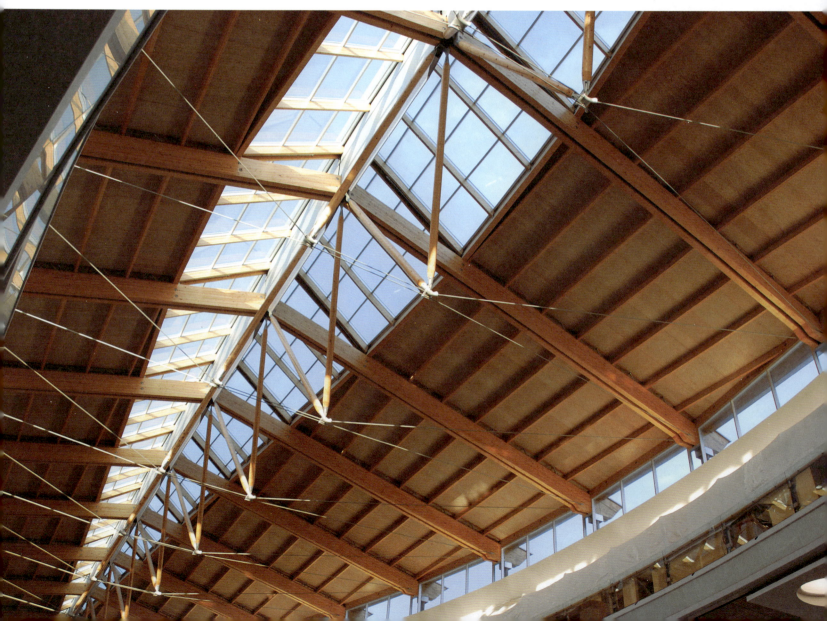

项目业主：ICBC 财产
建筑师：Bing Thom 建筑事务所
结构工程师：Fast + Epp
竣工时间：2003 年
建筑面积：7896 平方米
获奖情况：
杰出商业奖，素里新城市设计奖
杰出室内奖，素里新城市设计奖
引用奖，木结构设计奖
照明工程学会照明设计奖
MIPIM 评委会特别奖
不列颠哥伦比亚省建筑奖
不列颠哥伦比亚省建筑协会创新奖

Three distinct timber structural systems are used for the atrium, galleria, and facade of this combined shopping center, commercial office space, and university, giving warmth and expression to an otherwise concrete, steel, and glass building.

The atrium roof features a geometric wood space-frame constructed from 3,700 Douglas-fir peeler cores—making full use of a byproduct of the plywood industry. Varying clusters of Douglas fir logs, turned and tapered, branch from reinforced concrete columns. Upon entering the building, visitors' eyes are immediately drawn to the atrium's expressive web of three-dimensional timber, reminiscent of a child's Tinkertoy construction set. Large Parallel mullion columns support the atrium's lower glazing as well as the concrete canopy above it, something unique in a facade so large and complex.

三种不同的木结构系统被用于中庭、连廊和综合购物中心、商业办公空间和大学建筑的立面。在混凝土、钢和玻璃建筑之外，木结构传递出一种温暖的感觉。

中庭屋顶的特点是一个几何木材的空间框架，由3700根花旗松胶合板材为核心的构件组成，是胶合板产业副产品的充分利用。不同的花旗松原木杆件，呈圆锥形，在钢筋混凝土墩柱上形成分支。进入建筑后，参观者的视线立即被中庭富有表现力的三维木材网格所吸引，让人想起孩子的"万能工匠"建筑玩具。巨大的平行排布的竖向木柱支撑着中庭下部的玻璃，以及其上方的混凝土雨篷。这在一个如此庞大而复杂的立面形式中，是独一无二的。

The 2200-square-metre galleria roof—a free-form skeletal structure consisting of twenty individual three-dimensional composite timber-and-steel cable trusses—covers a serpentine-shaped, five-story-high vaulted space. The trusses are made with lightweight spruce glulam to minimize the crane loads.

All in all, the atrium, galleria, and facade designs not only push the boundaries of what's possible with wood, but serve as a civic statement, an illustration of the City of Surrey's official motto: the future lives here.

2200 平方米的连廊屋顶是一个形式自由的骨架式结构，其由 20 榀独立的三维复合木和钢索桁架组合而成。此结构之上是蜿蜒曲折的五层高的拱顶空间。桁架由轻质云杉胶合木制成，以减少起重机的负载。

总而言之，中庭、连廊和立面的设计不仅突破了木材的界限，同时也作为一种城市宣言，展示了素里市的官方格言：未来就在此处。

美国竖屏沃敏斯特总部办公楼
Vertical Screen Warminster Campus, USA

作者：贺雄岩，Shrey
翻译：张云川，Susan
校对：东鸿，蒋音成
摄影：杜红

Architect: Erdy McHenry Architecture, LLC
Engineering: The Harman Group & Shanghai Zhenyuan Timber Structures Engineering Co., Ltd.
Location: Southampton PA USA
Gross Floor Area: 3675 m²
Number of Floors: 1
Completion Date: 2010.8

项目设计：Erdy McHenry Architecture, LLC
结构设计：The Harman Group & 上海臻源木结构设计工程有限公司
项目地点：美国宾夕法尼亚州
建筑面积：3675 平方米
层数：单层
竣工时间：2010 年 8 月

This project is located in Southampton PA USA. The built-up area is about 38860 sqft(about 3675 square meters). The main structure is composed of (11) 24F-V5 curved southern pine glulam beams and more than three hundred and sixty glulam purlins. The curved beams have a span of 144 feet and are 44 feet height and 25 feet apart. The maximum overhang of wooden beams is 17 feet.

The bottom of the curved beams is exposed to the weather. In order to protect the beams from the damage by wind and rain, CCA anticorrosive treatment is applied to the bottom of the curved beams within 15 feet. The steel cables between curved beams supply the stability on the direction vertical to the curved beams. The project uses a total of about 440 m³ of American Douglas fir. glulam.

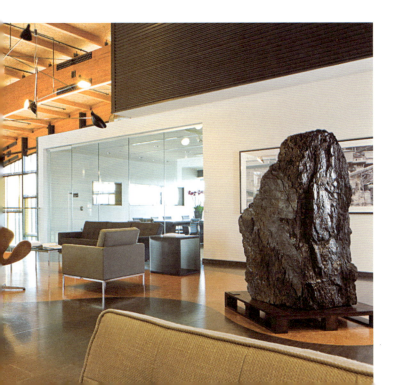

　　本工程是位于美国宾夕法尼亚州南安普顿的一综合办公项目，总建筑面积约 38860 平方英尺（3675 平方米），主体木结构由 11 榀 24F-V5 美国南方松胶合木曲梁及 360 余根胶合木檩条组成。曲梁跨度 144 英尺，高 44 英尺，间距 25 英尺，木结构最大悬挑 17 英尺。

　　曲梁底部暴露在室外，为了抵抗风雨对木材的损害，胶合木曲梁底 15 英尺高度内采用 CCA 防腐处理。曲梁之间的钢拉索为结构提供曲梁外的稳定性。工程共使用美国花旗松胶合木约 440 立方米。

186

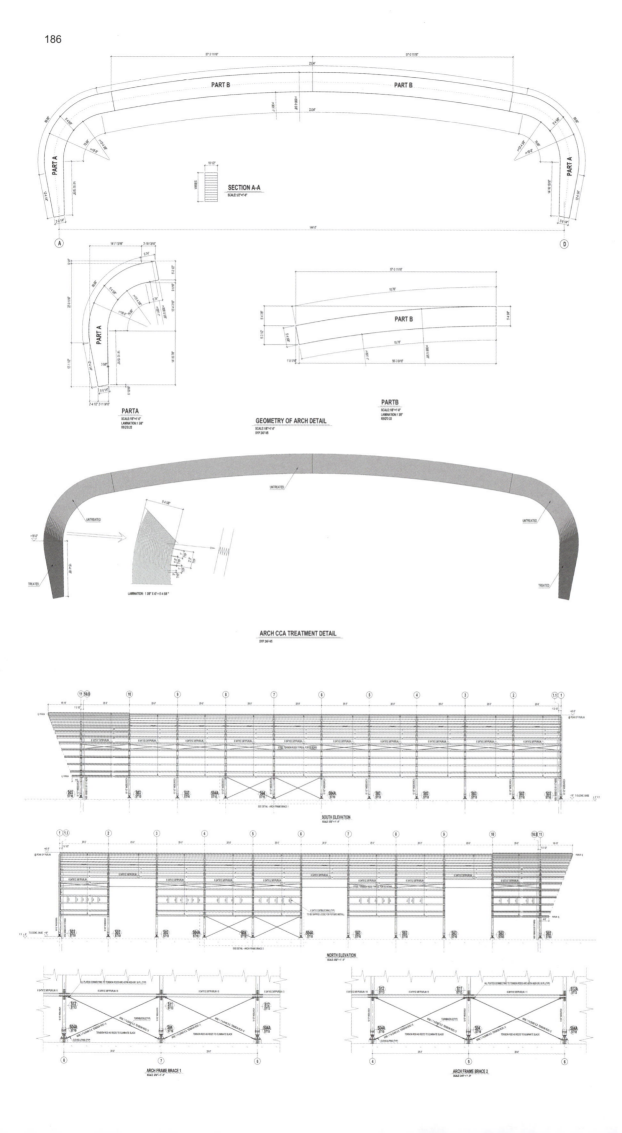

It took Shanghai Zhenyuan two months from the design to the completion of glulam fabrication and transportation to the construction site. In the next two months, the main wooden frame was erected.

The unique pattern and texture of the wood gives a warm and intimate feel, so it doesn't need extra decoration. The large open structure is very helpful for furnish and use. The large-span space allows owners to adjust and supplement the interior layout as needed. The building has a unique shape and is full of modernity.

The outdoor terrace had a good view, and is suitable as a social and resting place.

工程从设计到胶合木加工完成运抵施工现场共耗时两个月时间。在接下来的两个月内木结构主体框架施工完成。

木材特有的纹理和质感给人温暖亲切的感觉，表面也不需要额外的装饰。大跨度的空间方便业主可按需求对室内布局进行调整及补充。建筑拥有独特的外形，现代感十足。

室外平台拥有良好的视野，适合作为一个社交和休憩场所使用。

美国华盛顿水果生产公司总部办公楼

Washington Fruit & Produce Co.
Headquarters, USA

来源：grahambabaarchitects.com
翻译：蒋音成
校对：沈姗姗
摄影：Kevin Scott

The client, Washington fruit & produce, had a favorite barn that they wanted Graham Baba Architects to take their inspiration from. The solution was a building with simple geometry, an exposed structural frame and a natural patina, in a design that reflects the rural setting and a modern, scaled back aesthetic.

客户，华盛顿水果生产公司特别喜欢谷仓，他们希望 Graham Baba 建筑事务所能从中得到灵感。最终的方案是设计一栋有着简单几何外形的建筑：外露的结构框架、自然的光泽，在设计中体现田园风光，小而美。

The soil that was excavated to build the 1650 square meter office building was used to form rolling banks nearby, as a way to dampen the sound of the industrial buildings. An opening in the banks leads to the courtyard and entrance. What greets visitors is a 53-metre long building on one level, at most six meters tall, with an exposed glulam structure. There are no pillars on the inside, since the doubled and tied roof trusses extend through the façade, where they rest on crossed pairs of pillars.

1650 平方米的土壤被挖掘出来以建造办公楼；这些土壤被用于建造场地周边的起伏挡墙，以消除工业建筑产生的声音。挡墙中的一个开口，通向天井和建筑入口。欢迎游客的是 53 米长的一层建筑；包括裸露在外的胶合木结构，建造总高大约为 6 米。双层，且固定在一起的屋顶桁架延伸至建筑外立面，由交织成对的柱子支撑，所以这栋建筑内部没有柱子。

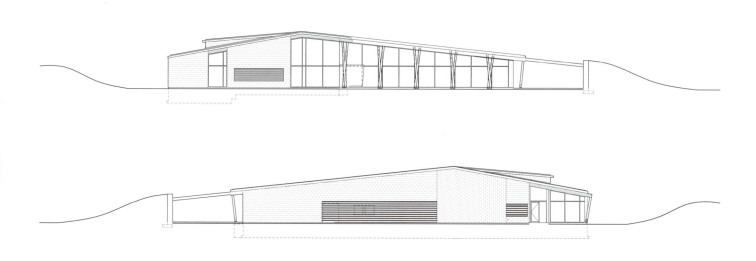

Internally, the 20-metre long roof trusses remain exposed. They all have the same centre spacing, which has cut the cost of manufacture and assembly. Their repetitive design also allows for a natural division of the space into different rooms. The open-plan layout is broken up by meeting rooms and other functions.

在建筑内部，20米长的屋顶桁架仍然裸露在外。他们之间都有相同的中心间隔，这样的方式减少了制造和组装的费用。这种重复的设计把空间自然地分割成不同的房间。这种开敞式平面布局由会议室和其他功能空间所分隔。

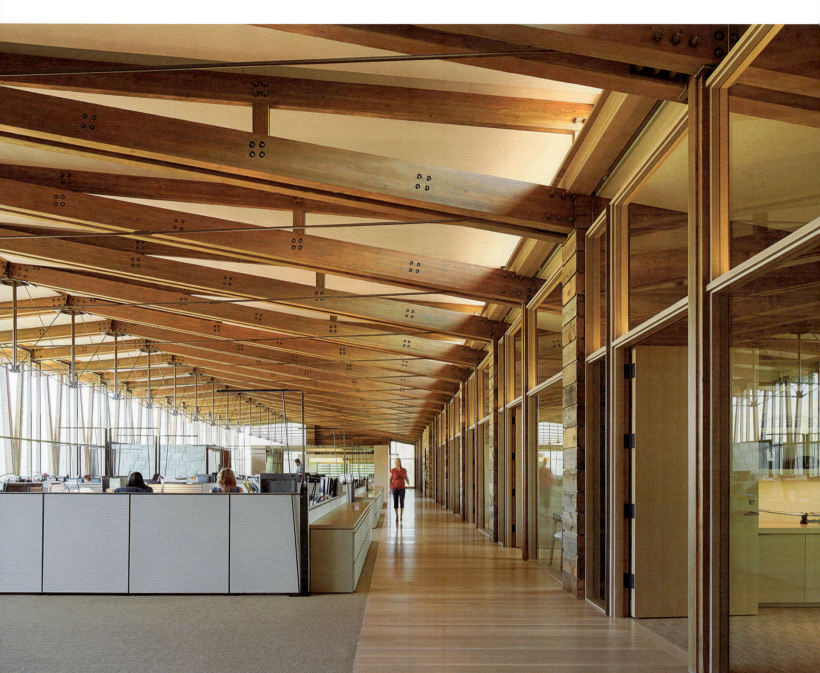

荷兰库肯霍夫公园中心
Keukenhof Flower Garden Center, Netherlands

来源：mecanoo.nl
翻译：蒋音成
校对：沈姗姗
摄影：Mecanoo architekten

Visitors to keukenhof's world-famous gardens are now welcomed by a standout entrance. Interwoven glulam triangles form an impressive roof over two volumes. Like five gigantic steps, the roof climbs upwards to create drama and level changes. The roof envelops and clings to and over the two volumes. One of these looks to have just a single level, but the upper floor is concealed by the roof and draws natural daylight in through lantern lights instead.

参观世界著名的库肯霍夫公园的游客们，现正受到这个非常显眼的入口大厅的欢迎。交织的胶合木三角形组件，构成了一个让人印象深刻的、横跨两个建筑体之间的屋顶。整个屋顶，如同5级巨大的楼梯，每一级屋顶向上攀升，以营造戏剧感和变化感。整个屋顶覆于两个建筑体之上。其中一个建筑体，看上去仅仅为一层，它的顶层隐于木屋顶之中，自然光照射进入，代替了人工照明。

The large roof also forms two plazas. The one outside the park welcomes visitors and is designed to guide them from the car park to the entrance. Inside the park, the roof floats above the entrances to the restaurant and the shop. The glazed façade can be opened on hot days and offers visitors a panoramic view of the park.

巨大的屋顶也构成了两个广场。公园外侧的那个广场，用于迎接游客，为游客指明从停车场到公园入口的方向。公园内侧的那个广场，屋顶漂浮在去往餐厅和纪念品商店的入口上方。玻璃外墙可以在炎热气候下打开，也为游客们提供公园的全景。

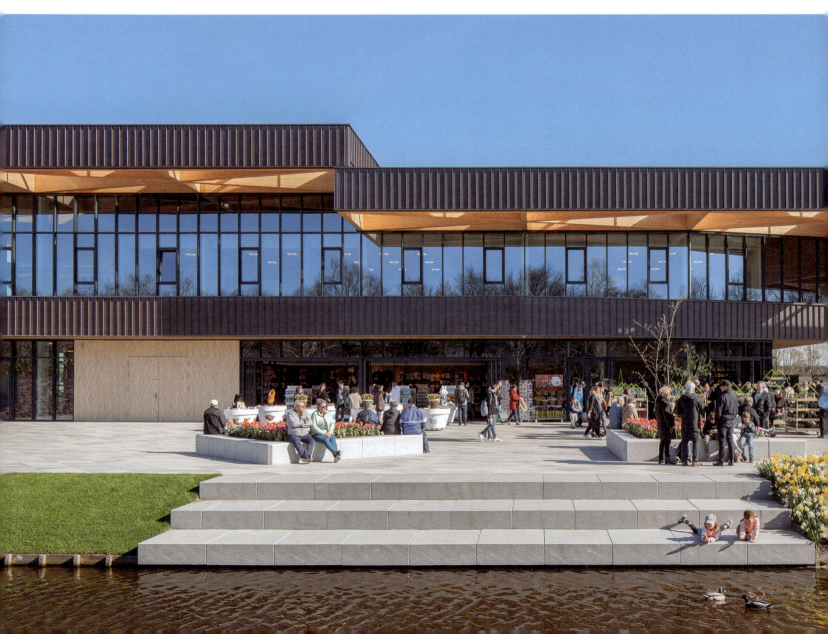

The large glulam beams, arranged in triangles, are clad in copper. Glass covers the openings in the triangles, providing shelter from the elements while also allowing an attractive play of light on the ground. Mecanoo Architecten is the firm behind the 3200 square meter project, which also includes 1.3 hectares of landscaping.

巨大的胶合木梁（构成三角形形式）屋顶四周悬挂铜板装饰。玻璃被放置在三角形的开口处，起到了避风雨的作用，同时，让阳光投射到地面上形成吸引人的光影效果。Mecanoo 建筑师事务所承接了这个 3200 平方米的项目。这个项目也包括了 1.3 公顷的园林美化工程。

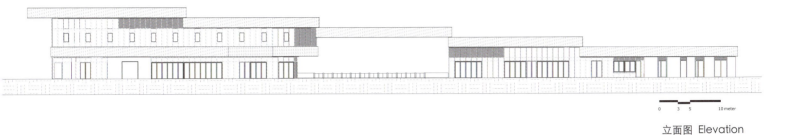

立面图 Elevation

Start of Planning: 2015
Completion: 2020 (expected)
Site area: 182 hectares
Designer: Delugan Meissl Associated Architects

Taiyuan Botanical Garden

The project introduces a vibrant artificial landscape with an attractive topography with mountains and hills, wild and elegant vegetation, lakes and waterfalls, paths and buildings. Nature and architecture communicate here in a harmonious way – the green space assuming a leading role.

The architectural concept is based on the already existing landscape plan and balances natural landscape, architecture, technological and ecological solutions.

Entrance Building
The entrance building, connected to the street and a big parking lot, plays a dual role: it invites the urban landscape into the garden and at the same time lets the natural landscape interact with the built environment. A cantilevered viewing platform above the water level leads the visitors to the center of the Botanical Garden.

Wood Gridded Glass Domes - Green House
In the project of Taiyuan Botanical Garden, the heart of the attraction will be three vaulted glasshouses fostering distinct environments for plants that thrive in different climates. Tropical, desert and aquatic plants will be housed in a trio of gridded glass domes at the center of these botanical gardens. Two of the three domes are designed to house species from arid or tropical environments, while the third – half-submerged in a lake – is for aquatic varieties.

Bonsai Museum
Another highlight of the Botanical Garden is the Bonsai Museum, designed as a rotated bowl integrated into the park's topography, a welcoming arena for the plants' display.

Restaurant Tea-House
The Restaurant is designed in resemblance to the Chinese Traditional Temples, evoking the wooden structures as its design premise. A shifted grid made from stacked timber beams was established as the main conductor of a very attractive and atmospheric space. The restaurants geometry detaches itself from it, extending opening up to the lake.

Research Centre
The Research Centre houses laboratories, studios, office spaces, workshop and meeting rooms, lecture halls as well as a library. In order to maximize its function, the program is divided into separate buildings, connected by an interior path along the ground floor.

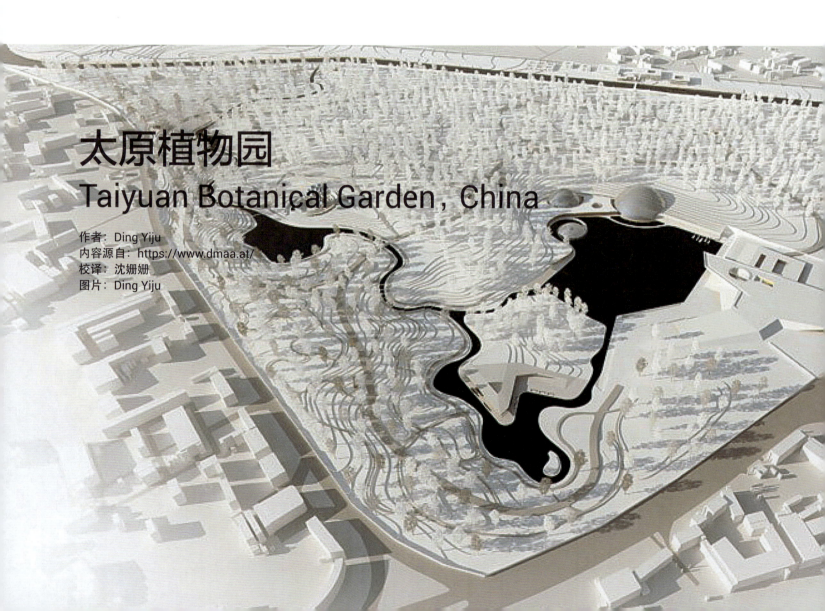

太原植物园
Taiyuan Botanical Garden, China

作者：Ding Yiju
内容源自：https://www.dmaa.at/
校译：沈姗姗
图片：Ding Yiju

规划设计：2015 年
竣工：2020 年（预计）
基地面积：182 公顷
设计单位：奥地利维也纳德隆岗梅西合伙人建筑设计事务所

太原植物园建筑设计

该项目引入了一个充满活力的人工景观。山、植被，湖泊和瀑布，道路和建筑物在这里自然交融。自然和建筑在这里和谐地交流。绿色空间起主导作用。

建筑理念基于现有的景观规划，通过建筑平衡自然景观、建筑、技术和生态之间的关系。

入口综合服务建筑

入口建筑起到了连接城市街道和大型停车场的双重作用，它将城市景观引入公园。同时建成环境与自然景观相结合，构成密不可分的一个整体。跨水而建的大悬挑观景平台将游客引导至植物园的核心区域。游客不但可以在上欣赏公园的美丽景色，同时可以通过平台斜切的大圆洞进入整个公园。

木结构温室

在太原植物园项目中，最吸引人的地方是三个拥有完全不同自然环境的温室。在这里，热带、沙漠和水生植物将被安置在植物园中心的三个木结构玻璃穹顶中。

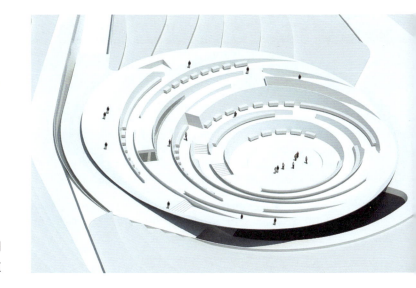

盆景博物馆

依地形而建。设计成一个螺旋的碗状，与公园的缓坡地形融为一体，是盆景植物的良好展示空间。

餐厅 & 茶室

餐厅的设计灵感来源于中国传统斗栱结构。运用现代的结构手法，由堆叠的木梁构成的空间木架结构被建立成一个非常吸引人的大气空间。餐厅的几何结构面向湖面延伸形成美丽的倒影。

科研中心

研究中心设有实验室、工作室、办公室、会议室、演讲厅以及图书馆。为了最大限度地发挥它的功能，他们被划分成独立的建筑物，同时通过沿着底层的内部路径连接起来。

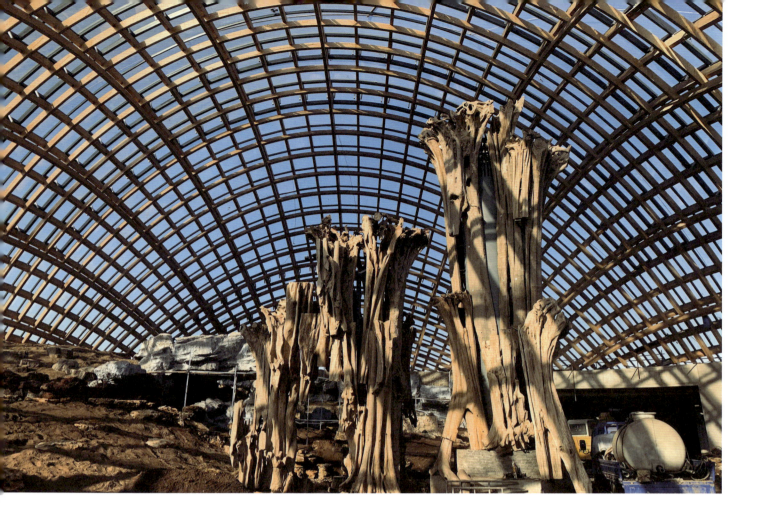

Greenhouse

The Greenhouse represents the centrepiece of the new Botanical Garden. It is composed of three domes destined to accommodate plants of different climates, together reinventing the silhouette of the Garden. Two of the three domes accommodate the pavilions for tropical and desert plants.

They are integrated into the exterior topography, facing south for maximum exposure during the summer and winter, potentiating an energetically and structurally well though-out solution.

The third dome – the aquatic plants– is designed as a stand-alone piece on the lake. The choice of materials follows the subject-matter, using natural elements to create the atmosphere.

温室

温室是新植物园的中心。它由三个圆顶组成，旨在容纳不同气候的植物，共同重塑花园的轮廓。

有着防寒、加温和透光功能的温室不仅是建筑师为植物精心打造的庇护所，而且也是参观游客的室外桃源。游客在室外感受植物四季变化的同时，也可以在温室内体验世界各地植物的异域风情。

静静的湖边镶嵌了三颗形态各异的"露珠"，分别是珍稀植物馆，沙漠植物馆和热带植物馆。而三座异球体的规模由小到大，直径分别为43米、54米、88米。

温室的最突出之处是将木结构的承重性能和审美性做完美的结合，木结构温室也充分体现出在一个享有中国古建筑博物馆之美誉的山西的传承和发展。

平面图 Plan

挪威阿尔嘉德的新木质教堂
Ålgård's new wooden church, Norway

作者：Erik Bredhe
翻译：蒋音成
校对：沈姗姗
摄影：Hundven-Clements photography

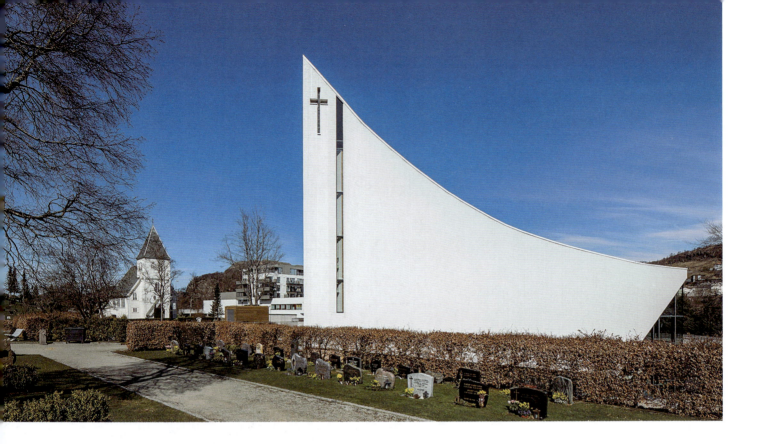

LINK Arkitektur has 14 offices and over 350 staff in Norway, Sweden and Denmark. Each office has developed its own specialist areas, based on its local knowledge.
Client: Gjesdal Church Council
Contractors: Faber Bygg (main contractor)、Kruse Smith (concrete)、Rubicon (glass façades)
Cost: Approx. NOK 50 million

Church In Two Shapes

The church is essentially built around two shapes: a triangle and a square, both of which are strongly symbolic in Christianity. The triangle represents the Trinity and the square stands for the four corners of the globe.

The building is set into the ground to create space for the lower level. This maximises the distance from the place of worship, ensuring that the activities down below do not disturb the people in the church above.

LINK Arkitektur 有 14 处办公室和超过 350 名员工分布在挪威、瑞典和丹麦。基于本地的情况，每个办公室都发展了自己的专攻方向。

客户：Gjesdal Church Council

承包商：Faber Bygg（主承包商）、Kruse Smith（混凝土）、Rubicon（玻璃立面）

花费：大约 5000 万挪威币

两种形状的教堂

教堂基本上是围绕着两个形状建造的：一个三角形和一个正方形，这两个形状在基督教中都具有很强的象征意义。三角形代表（圣父、圣子及圣灵的）三位一体，正方形代表地球的四个角。

建筑被设置在地面里，创造低层空间。这最大限度地远离礼拜场所，确保这些活动不会打扰上面教堂里的人们。

Ålgård's new wooden church — a cultural meeting place with open arms and windows to the heavens.

A misty valley in the evening sun. Trees and mountains as far as the eye can see. A river that flows ever so gently down at the bottom. And suddenly: an American football flies through the air, over the mountains, high above the spruce forest, and finally lands in a boy's arms, far out on a rocky ledge. The popular YouTube videos 'Kickalicious 1 & 2' from a couple of years back put Norway's Ålgård and the surrounding area on the map — but above all it took the star of the clips, Håvard Rugland, to the NFL and a career in American football.

Now Ålgård and the magnificent landscape in south-west Norway once again have the eyes of the world on them. The newly built Ålgård Church is an architectural masterpiece that has become a beautifully low-key local landmark since it opened in spring 2015. From a distance, the church looks like an enormous tent, with the white walls and sloping roof giving the impression of a light, soft canvas raised up on posts. At one end of the church, the entrance and steps symbolically face towards the village. At the other end, the church sweeps up in a gentle curve towards the sky. In one dynamic movement, the church rises up out of the landscape, like a natural continuation of Ålgård's hills and valleys. The architectural practice behind the project is Norway's LINK Arkitektur.

"We wanted to make the church feel like a tent, in order to create an open and inviting impression. It also symbolises Moses and the Israelites wandering through the desert. The idea was to create a modern interpretation of a traditional church." says chief architect René de Groot.

The church is not just for Christian services, it also serves as an assembly hall for the whole community. The building contains offices, group rooms, a café, space for ballgames and dancing, rehearsal rooms and so on. The place of worship itself is on the upper floor, while all the other activities are concentrated on the lower floor. This proved something of a design challenge, since the different activities had to work together, while still being able to be kept apart.

"There are separate entrances to the church and to the lower floor. The steps up to the church itself clearly mark it out as the 'main space' and make it easy to find. But however you look at it, everything is housed in one and the same building. Our greatest challenge was therefore to find a solution that contained all the functions of a modern cultural building, while also ensuring it works as, and feels like, a church. We were looking for a design that had a perfect balance between sacred and user-friendly." says René de Groot.

The shape of the church is square, with a diagonal axis. The design was driven by the client's brief for a compact, functional and economical solution. In using a square base, with the lower floor partially dug into the ground, it was possible to create a smart solution with short distances and reduced costs for external walls. Using wood as the construction material, both internally and externally, felt like a natural choice to the architects.

阿尔嘉德的新木质教堂——一个新的文化会议中心，以其张开的手臂和窗，通向天堂。

夕阳下的薄雾笼罩的山谷。一望无际的树木和山脉。一条在河底缓缓流淌的河流。突然：一个美式足球在空中飞过，越过高山，高高地飞过云杉林，最后落在一个岩石峭壁上的男孩怀里。几年前，YouTube 上的热门视频《Kickalicious 1 & 2》让挪威的阿尔嘉德及其周边地区闻名于世，但最重要的是，它让视频中的明星 Havard Rugland 进入了美国橄榄球联盟，并在美国橄榄球界开创了自己的职业生涯。

现在，阿尔嘉德和挪威西南部壮丽的景色再一次吸引了全世界的目光。新建的阿尔嘉德教堂是一件建筑杰作，自 2015 年春季开放以来，它已经成为当地一个美丽而低调的地标。从远处看，教堂就像一个巨大的帐篷，白色的墙壁和倾斜的屋顶给人一种轻盈、柔软的帆布搭在柱子上的感觉。在教堂的一端，入口和台阶象征性地朝向村庄。在另一端，教堂以柔和的弧线向天空延伸。从其动势来看，教堂从景观中升起，就像阿尔嘉德丘陵和山谷的自然延续。该项目背后的建筑设计公司是挪威的 LINK Arkitektur 建筑师事务所。

首席建筑师 Rene de Groot 说："我们想让教堂感觉像一个帐篷，以创造一个开放和好客的印象。它也象征着摩西

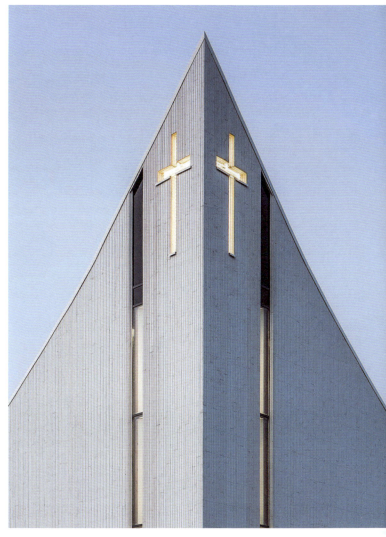

和以色列人在沙漠中漫步。我们的想法是创造一个基于传统教堂的现代诠释。"

教堂不仅仅是为基督教服务的，它还是整个社区的集会场所。该建筑包括办公室、集体活动室、咖啡厅、球类和舞蹈场地、排练室等。做礼拜的地方在上层，其他活动都集中在低层。这在某种程度上是一个设计上的挑战，因为不同的活动必须共处一室，同时仍然能够保持相对独立。

Rene de Groot 说："教堂和建筑的底层有各自的独立出入口。通往教堂的台阶清楚地标示出它是一个'主要空间'，很容易找到。但是无论你怎么看，所有的空间都在同一栋楼里。因此，我们最大的挑战是找到一个方案，可以包含现代文化建筑所有功能，同时确保它像教堂一样运作，而且感觉也像一个教堂。我们正在寻找一个在神圣和用户友好之间达到完美平衡的设计方案。"

教堂的形状呈斜轴方向的正方形。该设计取自客户关于集约、功能性强和经济节省的想法。在使用方形底座时，低楼层部分挖入地下，这创造出一个可以缩短外墙尺寸、降低外墙成本的好方案。无论是建筑内部还是外部，使用木材作为建造材料，对建筑师来说都是一个自然的选择。

"Wood is a warm and humane material, which fitted in perfectly with the function of Ålgård Church. There is also a Nordic tradition of building wooden churches, including the old church nearby, which is made of wood." explains René de Groot.

The façade, roof and structural glulam beams are spruce. The façade is painted to protect it against the elements, and is treated with a fire retardant finish. Internally, the church is fitted with sprinklers for additional fire safety. The floor of the church is laid with beautiful oak. The same wood is also used for the vertical elements that combine with white painted plasterboard to make up the interior walls. The glulam beams that have been left visible in the ceiling are one of the first details to strike visitors to the church. Together, they create a triangular pattern reminiscent of a beehive. The design has structural benefits, as well as an aesthetic and symbolic function. Arranging the beams in this way creates a rigid structure that is capable of carrying the roof. But equally important is the symbolism of the pattern, since in Christianity the triangle represents the Trinity. It is also a nod to what the vaults inside traditional churches usually look like. The pattern is further accentuated by the nine triangular windows placed in the roof.

"The building has no regular windows. The ones up in the ceiling give the impression of lifting up the 'canvas' and carefully letting light into different parts of the space. These windows to the heavens create contact between the inside and the outside. Additional daylight is admitted behind the altar, through the glass artwork standing there." says René.

The church has proved a big hit with the locals. René de Groot reports that many of those who don't usually go to church have been to this one. At the official opening earlier in the spring, the church was packed with over 600 inquisitive visitors.

They squeezed into the building to listen to the Bishop of Stavanger, Erling Pettersen, giving the church his blessing. His words made the local residents glow with pride, and LINK Arkitektur was able to relax, safe in the knowledge that they had created a modern, inclusive place for people of faith and those with none:

"This is one of the most beautiful churches I've ever been in. From now on, I'm going to call it the Church of the Open Arms."

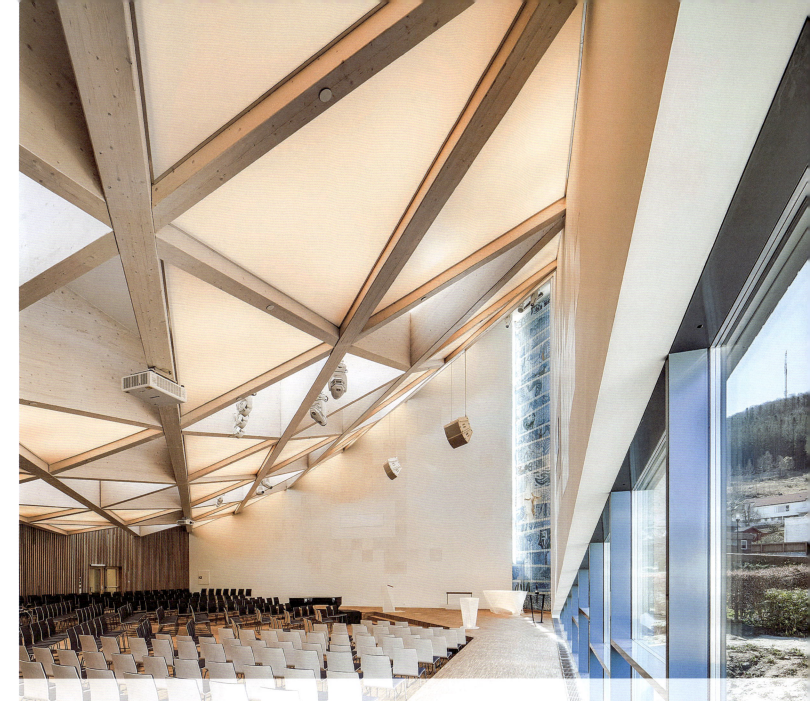

"木材是一种温暖而人性化的材料,它与阿尔嘉德教堂的功能完美契合。北欧也有建造木制教堂的传统,包括附近的老教堂都是用木头建造的。"Rene de Groot 解释说。

立面、屋顶和结构胶合梁均采用云杉。外立面被涂上油漆,以保护它免受外界因素的影响,并进行了防火处理。教堂内部安装了喷淋装置,作为额外的消防安全措施。教堂的地板是用漂亮的橡木铺的。同样的木材也用于竖向面板,与白色石膏板相结合,制作成内墙。天花板上外露的胶合木梁是教堂吸引游客关注的首批细节之一。他们一起创造了一个三角形的图案,让人想起蜂巢。

该设计不仅具有结构上的优势,而且具有美感和象征功能。以这种方式布置的梁结构,创造了一个能够承载屋顶的刚性结构。但同样重要的是这个图案形式的象征意义,因为在基督教中三角形代表(圣父、圣子及圣灵的)三位一体。在传统教堂中拱顶也有相同的象征意义。屋顶上的九个三角形窗户则进一步加强了这种图案形式。

"这栋建筑没有常规的窗户。天花板上的窗随着'帆布帐篷'屋顶向上翻起,小心翼翼地让光线进入空间的不同部分。这些通向天空的窗户,使建筑的内部和外部空间联系起来。额外的日光通过这些玻璃艺术品,从祭坛后方进入室内。"Rene 说。

事实证明,这座教堂受到了当地人的热烈欢迎。据 Rene De Groot 反映,很多平时不去教堂的人都去过这个教堂。在今年早春的正式开幕仪式上,教堂里挤满了 600 多名感兴趣的访客。

他们挤进教堂,聆听斯塔万格主教 Erling Pettersen 为教堂祈福。他的话让当地居民感到肃然起敬。LINK Arkitektur 建筑事务所才得以放松,因为他知道,他们为有信仰和没有信仰的人创造了一个既现代化又有包容性的场所:

"这是我去过的最美的教堂之一。从现在起,我要把它叫作张开双臂的教堂。"

Location: Abbotsford, British Columbia
Size: 4645 m²
Completion: November 2017
Architects: Keystone Architecture
Structural engineer: StructureCraft Builders Inc.
General contractor: StructureCraft Builders Inc.
Engineered wood supplier/fabricator: StructureCraft Builders Inc.
Project owner: StructureCraft Builders Inc.

Awards:
Prefabricated Structural Wood, 2018 WOOD DESIGN AWARDS
The Award for Outstanding Value, Structural Awards 2018

Project Overview

StructureCraft's new manufacturing facility demonstrates the possibilities for using a wood superstructure to construct an industrial building. Using engineered wood in prefabricated wall and roof panels, the building was erected in just five days. Traditionally, these types of industrial buildings have been built with tilt-up concrete walls and steel roof structures, but StructureCraft, a mass timber design and production company, took the opportunity to design a new industrial building prototype out of wood for their own manufacturing plant. The new facility makes them the first North American company to manufacture dowel laminated timber, and the project showcases the ability to efficiently and economically construct industrial buildings from wood.

The building, which includes 4000 square meters of manufacturing and about 650 square meters of office space, was designed and built as a demountable structure, with modular wood wall and roof panels. Modular configuration enabled the almost unheard-of construction speed of five days; the erection of the timber superstructure commenced on a Monday, and the crew had the entire building, walls and roof, erected by Friday evening of the same week. The building was also cost competitive, costing about the same as a concrete tilt-up building but fully insulated, and with the aesthetics of exposed wood, providing a more pleasing work environment than a typical warehouse, further contributing to the overall health and well-being of the workforce.

The strength of wood is demonstrated by the fact that crane rails supporting the heavy-duty manufacturing equipment are directly attached to the wood columns and walls leaving no need to add a steel support structure. Since the demountable wood structure is panelized and screwed into place, it is also possible to take the building apart and extend it to accommodate future facility expansion.

Wood Use

StructureCraft used much of their own engineered wood products, manufactured and/or fabricated at their former plant in Delta for this project. Glulam beams and columns, supplied by others, form the structural frame of the new facility, and special architectural quality nail laminated timber (NLT) panels were used for the roof of the office portion of the building. They also used plywood sheathed NLT panels for the primary shear walls.

For the manufacturing facility, the main roof structure was framed by glulam beams with purlins and layers of plywood on top to form panels 3.7 meters wide by 19 meters long. To manage lateral loads like wind, several layers of plywood were added. The multiple layers of plywood created a large structural diaphragm that distributes lateral loads to the four exterior walls and eliminates the need for any internal bracing in the building.

To form the fully-insulated, prefabricated wall panels, laminated strand lumber was used for the wall studs, spaced 0.6 meters on center. After adding insulation, plywood was fastened to both sides of the tall wall studs to create wall panels 3.7 meters wide by 10 meters tall. These wall panels were lifted and then dropped into place by crane on-site. The office floors and roof were built with NLT incorporating a special profile in each board containing acoustically absorptive material. The NLT used on the office built by hand, was created to showcase the possibilities of the similar but more advanced, dowel laminated timber to be machine manufactured in the new plant.

项目地点：阿伯兹福德，不列颠哥伦比亚省
建筑面积：4645 平方米
竣工时间：2017 年 11 月
建筑设计师：Keystone Architecture 建筑事务所
结构工程师：StructureCraft Builders Inc.
总包：StructureCraft Builders Inc.
工程木供应商：StructureCraft Builders Inc.
项目业主：StructureCraft Builders Inc.
获奖情况：2018 年装配式木结构建筑奖，木材设计奖
2018 年杰出价值奖，结构设计奖

加拿大"结构工坊"工厂
Structurecraft Manufacturing Facility, Canada

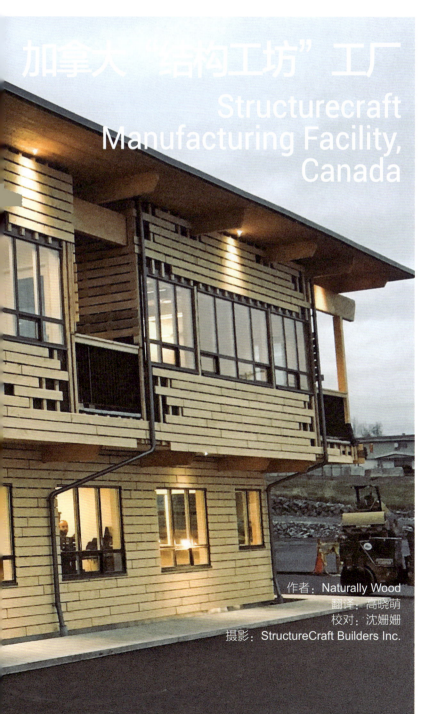

作者：Naturally Wood
翻译：高晓萌
校对：沈姗姗
摄影：StructureCraft Builders Inc.

项目概况

"结构工坊"的新生产设施展示了运用木上盖结构来建造工业建筑的可能性。运用工程木来建造预制墙体和屋顶板，建一栋建筑只需要 5 天。通常来说，这类工业建筑往往使用预制混凝土墙和钢结构屋顶来建造，但是"结构工坊"，一个大规模的木材设计和生产公司，利用这个机会为他们自己的生产工厂设计了一个新型的用木头制造的工业建筑样板。新设施使他们成为北美第一家生产木钉连接胶合木的厂家，这个项目也展示了木材高效经济地建造工业建筑的能力。

此建筑包括了 4000 平方米的生产空间和大约 650 平方米的办公空间。此建筑被设计和建造成可拆卸的结构，并使用了模块化的木墙和屋面板。模块化的构造使建造速度达到了闻所未闻的五天完成；建造上部木结构工作从周一开始，整栋建筑、墙和屋面在同一周的周五晚上被建造完成。建筑造价也很有竞争性，成本和预制混凝土建筑差不多，但是该建筑在预制阶段就完成了保温，且展示了裸露的木材美学，提供了一个比传统仓库更愉快的工作环境，进一步促进了员工的整体健康和幸福感。

起重机轨道支撑的重型生产设备直接连接在木柱和墙体上，体现了木头的强度，并且不需要增加钢结构支撑。因为可拆卸的木结构是用木板拼装和用螺丝固定的，所以也可以将建筑拆分并加以扩展，以适应未来设施的发展扩张。

木材使用

"结构工坊"工厂运用了许多他们自己的工程木产品，这些产品都是在 Delta 的原工厂生产和预制的。其他供应商提供的胶合木梁柱组成了新设施的结构，办公区部分的屋面使用了特殊建筑质量标准的钉连接胶合木（NLT）。同时也使用了胶合板覆盖的钉连接胶合木建造主要的剪力墙。

对于一个生产设施来说，主要的屋面结构是由带檩条的胶合木梁搭建，上覆多层胶合板，这些多层胶合板的尺寸为 3.7 米宽，19 米长。为了控制横向荷载，比如风荷载，增加了几层胶合板。这些多层胶合板形成了一个大型的结构横隔梁，将横向荷载分配到四个外墙上，消除了建筑内任何内部支撑的需要。

为了建造完整的保温预制木墙，预制墙体的框架柱使用了层叠木片胶合木（LSL），间距为中对中 0.6 米。在添加保温材料后，胶合板被固定在墙柱的两侧，形成了 3.7 米宽，10 米高的预制墙板。这些预制墙板被起重机吊起，放置在指定位置上。办公室的楼板和屋面板用钉连接胶合木建造，在每一块板内都包含了吸声材料。用于办公室的钉连接胶合木是手工制作的，这是为了展示一种形式类似但更先进的木钉连接胶合木在新工厂中进行机械生产的可能性。

奥地利菲沙门德物流中心
Logistics Center in Fischamend, Austria

作者：Erik Bredhe
翻译：蒋音成
校对：沈姗姗
摄影：Walter Ebenhofer

Poppe-Prehal Architekten in Steyr, Austria. Founded in 2000 by Helmut Poppe and Andreas Prehal. Works on public and commercial buildings, individual houses and whole residential developments. Focus on energy-efficiency, usability and aesthetics. Several of its large, ecologically-driven wood construction projects have won awards.

Client: ATL Immobilienverwaltung.
Structural engineer: Wiehag Timber construction (wood), Held & Francke Bauges (other).
Gross area: 12250 square meters.

Function and aesthetics in glulam and larch meet future needs thanks to meticulous planning.

For the opening of Cargo-Partner's newly built logistics center next to Vienna Airport, the traditional ribbon cutting ceremony was replaced with company CEO Stefan Krautner using a chainsaw to cut up a thick log. This was an appropriately symbolic choice considering the large amount of wood that has been used – a total of 4200 cubic meters of environmentally certified spruce and larch that was grown, harvested and sawn in Bavaria and Austria – and which now stores that same amount of carbon dioxide.

Poppe-Prehal Architekten 建筑事务所位于奥地利斯太尔，由 Helmut Poppe 和 Andreas Prehal 创立于 2000 年。该事务所主要从事公共和商业建筑设计、私人住宅和整个住宅区规划项目，注重节能设计、可用性及美观性。该事务所的多个大型以生态绿色主导的木结构建筑项目获得奖项。

客户：ATL Immobilienverwaltung
结构工程师：Wiehag Timber construction (wood) 和 Held & Francke Bauges (other)
总面积：12250 平方米

在精心策划之下，胶合木和落叶松木的功能和美学满足了未来的需要。

在 Cargo-Partner 货运公司新建的位于维也纳机场旁边的物流中心开幕式上，传统的剪彩仪式被公司首席执行官斯特凡·克劳特纳（Stefan Krautner）用电锯切开一根粗壮的圆木取而代之。这个是合适的象征性的选择，因为这个物流中心的建造使用了大量的木材——总共用了 4200 立方米的环保认证的云杉和落叶松（在巴伐利亚和奥地利地区生长、成熟、砍伐），且现在存储着等量的二氧化碳。

Construction of the enormous complex — 109m × 104m × 20m — took exactly 365 days from the first groundbreaking ceremony to the point when the 35 employees were able to get started in their new workplace.

With the steady growth in e-commerce and rising demand for the company's services, it became increasingly important for Cargo-Partner to expand its warehouse space and continue to embrace modern information technology in order to avoid bottlenecks in the logistical flow.

The job of producing a proposal for a modern and energy-efficient logistics center with the least possible environmental footprint went to Austria's Poppe-Prehal Architekten. During the short and efficient planning process, the client and the architect quickly agreed to build the whole thing in wood, on a foundation of concrete.

"High-tech logistics in a sustainable package" is how Stefan Krautner sums it up.

建造这座巨大的建筑群——109米×104米×20米——从第一次奠基仪式到35名员工能够在他们的新工作场地开始工作,整整花了365天。

随着电子商务的稳步发展和客户对公司服务需求的不断增加,Cargo-Partner货运公司为避免物流瓶颈,其扩大仓储空间并继续采用现代信息技术变得越来越重要。

奥地利Poppe-Prehal Architekten建筑事务所提出了一个既现代又节能的物流中心的方案,提出尽可能减少物流中心的环境影响。在短而有效的规划过程中,客户和建筑师迅速达成共识,即在混凝土的基础上建造整个木结构建筑。

Central factors in the project from the very beginning included climate considerations and long-term sustainability, with a side order of aesthetics and brand building. And although, at first glance, choosing wood as the main construction material appeared to be more expensive than traditional techniques using concrete and steel, with metal cladding on the walls, it turned out that the pricier alternative would make up the difference in cheaper maintenance after just a few years. It was also clear that a wooden structure not only brings lower carbon emissions, but is much easier to adapt to new needs that may arise in the future.

"And when the structure reaches the end of its life, instead of being carted off as hazardous waste, it can easily be dismantled and recycled as an eco-friendly energy source," states Verena Dolezal of Poppe-Prehal Architekten.

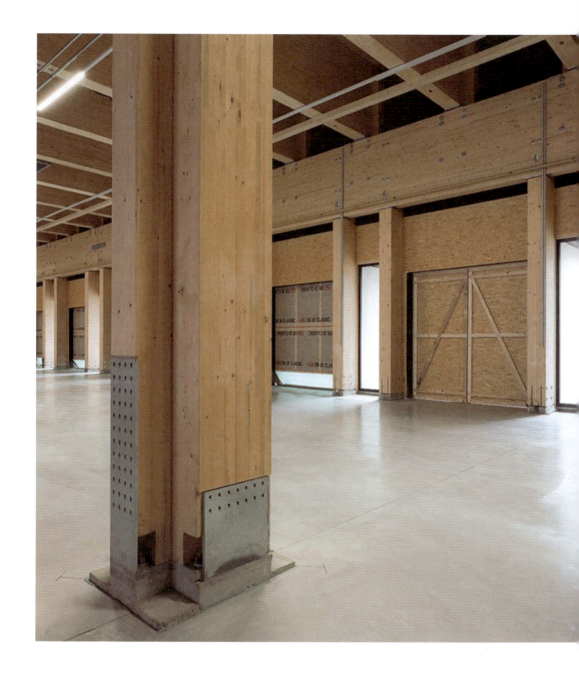

从一开始，项目的核心要素就包括气候因素和长期可持续性因素，其次是美学和企业品牌建设。虽然，乍一看，选择木材作为主要建筑材料似乎比使用混凝土和钢（墙面为金属挂板）的传统技术要贵；但事实证明，价格高的材料在今后几年的维护成本更低。另外很明显的是，木结构不仅能降低碳排放，而且更容易适应未来可能出现的新需求。

"当结构达到其使用寿命，它并不会被视作危险废物而运走，它可以很容易地拆卸，并作为一种环保能源来回收。" Poppe-Prehal Architekten 事务所的设计师 Verena Dolezal 说。

The structural frame of Cargo-Partner's new logistics center comprises a grid of studs, load-bearing glulam beams and 12 cruciform posts in glulam, measuring 1.45m × 1.45m × 16m, that hold the roof in place, with screwed plates and clamps at the nodes.

The majority of the volume is split into two levels, with space for 24500 pallets on the ground floor and an automated warehousing system for 30000 blue boxes containing spare parts, IT components and other smaller products on the upper floor. The top level's floor system combines three layers of wood paneling with a 10 centimeter thick wear layer of cast concrete.

The walls and ceiling are made from prefabricated elements that form a framework, insulated with 24 centimeter mineral wool and fire-retardant Agepan sheets. Externally, the entire building is clad in untreated larch. The roof is covered with a sheet of synthetic EPDM rubber, made from ethylene and propylene.

The technical challenges of a large-scale building in wood are considerable. There can be no compromise on the small tolerances required for the upper floor's computer-controlled warehousing system, despite the use of a living material that expands and contracts with changes in humidity and temperature.

"As well as stabilizing the ground beneath the posts, we calculated the dimensions of the beams and the distance we needed between the posts to avoid movements in the structure or a floor failing under a heavy load." explains Doris Klein, technician on the project management team at Poppe-Prehal.

Cargo-Partner 物流公司新物流中心的结构框架包括一个由墙体框架柱、承重胶合木梁和 12 根十字胶合木柱组成的网格，尺寸为 1.45m × 1.45m × 16m，将屋顶固定到位，节点处由带螺纹的钢板和夹具固定。

大部分空间分为两层，底层可容纳 24500 个货盘，上层是 30000 个装有备件的蓝盒、IT 组件和其他小型产品的自动化仓储系统。顶层的楼板系统由三层木板和 10 厘米厚的现浇混凝土面层组成。

墙体和天花板由预制构件装配出框架，内部填充 24 厘米矿棉和阻燃 Agepan 板进行隔热。从外部看，整座建筑都覆盖着未经处理的落叶松。屋顶覆盖着一层由乙烯和丙烯制成的合成三元乙丙橡胶板。

大型木结构建筑的技术挑战是相当大的。建筑使用的是随湿度和温度变化而膨胀和收缩的活性材料，但位于上层的计算机控制的仓储系统却不允许有较小公差。

"除了稳定柱子下面的地面外，我们还计算了横梁的尺寸和柱子之间的距离，以避免结构的位移或地板在重载下发生损坏。"Poppe-Prehal 建筑事务所的项目管理团队技术人员 Doris Klein 解释到。

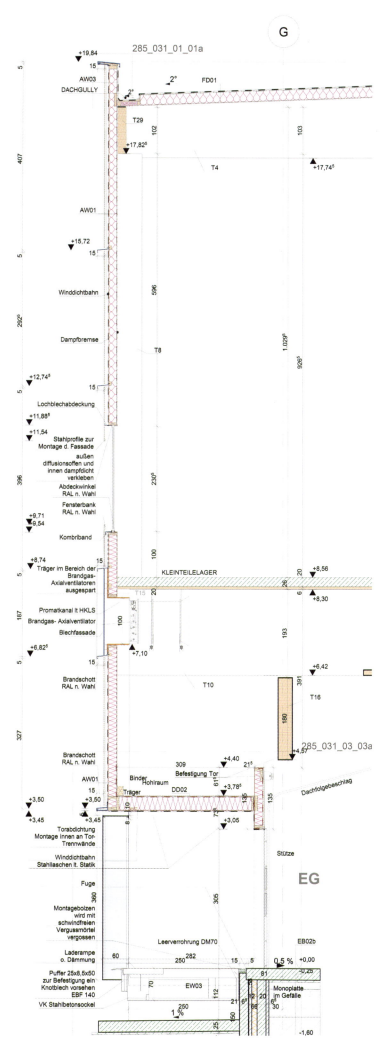

Early on in the process, an experienced structural engineer was brought in to help with the building's planning and design. Industrial architecture is usually associated with heavy time pressure, but the watchwords this time around were functionality, economy and aesthetics instead.

"There's a general misunderstanding that an attractive and sustainable industrial building has to be more expensive than an equivalent building in steel and concrete. The secret lies in meticulous planning down to the tiniest detail." says Helmut Poppe.

Wherever you look in the high building, wood is a striking feature, in the walls, ceilings, posts and beams. The only exception is a modest number of windows, whose light makes the wooden surfaces an even stronger presence.

The total floor space is 12250 square meters, with a volume equating to 6881 TEU containers, or 227760 cubic meters. The use of effective insulation has significantly reduced losses of heat and cold. Internally, the building maintains a cost-effective temperature of between 15 and 26 degrees Celsius, with a steady air humidity of 70 percent.

The building is fitted with 20 loading bays for incoming and outgoing goods. To protect the posts from stray forklifts, flexible bumpers have been embedded in the floor slab, And in the event of a fire, the wooden structure is designed to remain structurally sound for 60 minutes. A sprinkler system has also been installed throughout the building.

在这个过程的早期,一位经验丰富的结构工程师被请来帮助该建筑的规划和设计。工业建筑通常伴随着沉重的压力,但这次的口号是功能性、经济性和美学。

"大家有一个普遍的误解:一个有吸引力和可持续的工业建筑必须比同等的钢和混凝土建筑更昂贵。但秘诀在于周密的计划,直到最细微的细节。"Helmut Poppe 说。

无论你在哪个位置观看这座大楼,在墙体、天花板、柱子和横梁上,木材都是一个引人注目的特征。唯一的例外是数量不多的窗户,它们的光线使木质表面显得更加显眼。

总建筑面积为 12250 平方米,相当于 6881 个标准箱集装箱,即 227760 立方米。有效保温材料的使用大大减少了冷热损耗。在建筑内部,该建筑保持 15~26℃ 的经济而有效的室内温度,空气湿度稳定在 70%。

这栋建筑设有 20 个进出货物装卸区。为了保护木桩不受叉车带来的损伤,在地板上嵌入了弹性缓冲器。在发生火灾的情况下,木结构被设计为保持结构耐火 60 分钟。同时,整个大楼也安装了自动喷淋灭火系统。

法国喷气式飞机库
Garage for the Jetset, France

来源：comtevollenweider.fr
翻译：蒋音成
校对：沈姗姗
摄影：Luc Boegly

70000 aircraft take off and land every year at Cannes-Mandelieu Airport. Private planes packed with the jetset on their way to the French Riviera are not an uncommon sight. Many of them park in the airport's new hangar, designed by French architects Comte Vollenweider.

In the middle of the building, Cessna and Falcon private jets can be driven in and out through the gigantic 35 meter-wide door, while the back corners of the building, with their curvaceous lines, house office space for the airport's employees. Electromagnetic radiation from the airport's instruments made wood the most suitable choice for the structure.

每年有 70000 架飞机在康城机场起飞和降落。在去法国里维埃拉的路上，私人飞机和喷气式飞机是很常见的。它们中的不少飞机选择停在这个机场的新停机库，这个新停机场是由 Comte Vollenweider 公司设计的。

在这个建筑的中间，塞斯纳和菲尔肯私人喷气式飞机可以通过一扇 35 米宽的巨大门廊驶入和驶出；而在带着曲线设计建筑的后方，为机场的工作人员提供了办公空间。使用木材来建造结构，对机场设备的辐射来说，是最合适的材料。

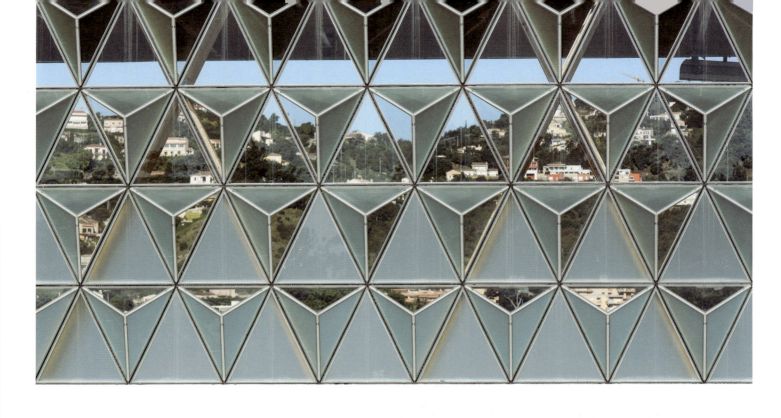

The hangar's load-bearing frame is made from pine glulam beams. The height of the beams varies between 560 and 790 millimeters. The V-shaped diagonals create an attractive and flexible design. The facade is dominated by the triangular windows, which have flat glass mounted in parallel with three-dimensional, pyramid shaped panes. Between these sit narrow ribs of pine.

这个停机场的承重框架是用松木胶合梁构建的；选取的胶合梁的高度从 560 毫米到 790 毫米不等。V 形的斜撑，营造了一个有魅力又灵活的设计。停机场的立面主要由三角形的窗户构成：平整的玻璃被平行地镶嵌在 3D 金字塔形的框架中。在玻璃之间，放置了狭窄的松木肋柱。

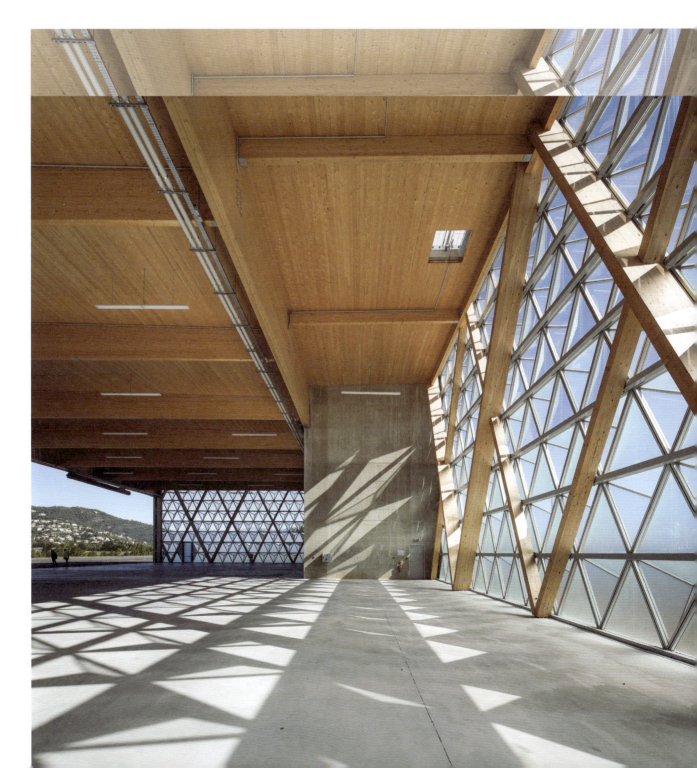

法国和瑞士的创意屋顶
Creative Roofs in France and Switzerland

作者：Mats Wigardt
翻译：蒋音成
校对：沈姗姗
摄影：Didier Boy de la Tour，Julien Lanoo，Leonardo Finotti，Shigeru Ban Architect，Dietrich Untertrifaller Architekten，neue Holzbau

Different construction systems in wood

Wooden structural frames with large spans are largely built using glulam nowadays. In its most common form, a glulam frame comprises simply supported beams resting on posts at either end. With glulam it is also possible to manufacture curved designs such as arches, grids, shells and so on.

Trusses are a common construction technique based on rods that are connected to form triangles, creating a very stable load-bearing structure where the rods are mostly subject only to normal forces, in contrast to beams, where the bending force dominates. With the help of trusses, it is possible to achieve very large structures that make efficient use of materials. Examples include the radio mast in Gliwice, Poland, and the Wildfire rollercoaster at Kolmården Wildlife Park in Sweden. The benefits of trusses include prefabrication, while one of the disadvantages is the number of sometimes complex nodes.

Grillage is a system where the beams are load-bearing in two, mutually perpendicular directions, which allows the structure to be lower in height. The system's benefits come to the fore in quadratic structures with an equal span in both directions.

Shell structures allow for more advanced designs with a diverse range of different roof shapes and large, post-free spaces.

木材的不同建造体系

大跨度木结构框架目前主要采用胶合木材料建造。其最常见的形式是由简支梁和在其两端起支撑作用的柱子所组成的胶合木框架结构。胶合木还可以做曲，如拱、曲面网格、壳体结构等。

桁架是一种常见的施工技术，受力杆件之间连接成三角形，从而创造出一种非常稳定的承重结构。这些杆件大多只受法线方向力的作用，而梁则主要受到弯曲应力。在桁架的帮助下，有效地利用材料建造出非常大尺寸的结构，是有可能实现的。例如，波兰格里韦斯的无线电桅杆，以及瑞典科尔马登野生动物园的野火过山车。桁架的优点是可预制，但复杂的节点是它的缺点之一。

格栅结构是一种梁在两个相互垂直方向上同时受力的结构系统，这使得其结构高度较低。该系统的优点在两个方向上跨度相等的二次结构中表现得尤为突出。

壳体结构可满足更高要求的设计，可建造不同的屋顶形状和大面积的无柱空间。

Advanced roof structures in wood have often served to surprise visitors and mark the importance of a building, a symbol of power and abundance for kings, emperors and spiritual leaders. Today there are creative roofs with load-bearing wooden components across much of the world.

There are many good reasons to choose wood when building a roof, not least its light weight, formability and cost-effectiveness.

When, in Germany in 1906, master carpenter Otto Hetzer patented his idea of gluing planed boards together to form curved beams, the result was what we now call glulam, a dimensionally stable material that is strong enough to be used across large spans.

The first big showcase of glulam's ability to push boundaries, its beauty, its high strength in relation to its own weight and its ability to optimise the properties of wood came at the World's Fair in Brussels in 1910, when large glulam arches were used to form the roof of the new Reichseisenbahnhalle.

先进的木结构屋顶常常让游客感到惊讶,并体现出建筑的重要性。对国王、皇帝和精神领袖来说,这是权力和富足的象征。今天,世界上很多地方都建造了由木构件承重的富有创造性的屋顶形式。

在建造屋顶时,有很多很好的理由来选择木材,尤其是它的轻质、可塑性和经济性。

在1906年的德国,木匠大师Otto Hetzer为刨花板粘合在一起形成弯曲梁的想法申请了专利,其结果就是我们现在所说的胶合木,一种尺寸稳定、强度足以跨越大跨度的材料。

1910年布鲁塞尔举行的世界博览会成为胶合木的第一次大展示:胶合木展示了其卓越之处,它的材料之美,它的高强度支撑力和自身重量以及优化木材性能的优点。同年,巨大弯曲胶合木被用来构建新Reichseisenbahnhalle的屋顶。

Swedish State Railways (SJ) was an early adopter of prefabricated glulam for the structural frames of halls and platforms all over Sweden. Elegant roofs built in locations such as Malmö, Stockholm, Sundsvall and Gothenburg remain in use to this day. And the wooden roof of a 1937 aircraft hangar in Västerås long held the record for the world's largest span at 55 metres!

Interest in creative and aesthetically attractive roof structures in glulam and CLT is now growing around the globe.

"With its strength and low weight, wood is superb for load-bearing, and often spectacular, structures with large spans." says Eric Borgström, structural engineer at Swedish Wood.

Japanese architect Shigeru Ban perhaps stands out when it comes to experimenting with climate-positive wooden designs for structural frames and roofs. In 2014 he was awarded the prestigious Pritzker Prize, often referred to as the Nobel Prize of the architectural world, for lauded projects such as Aspen Art Museum in the US, and the billowing wooden roof of the Centre Pompidou-Metz in France, not to mention his temporary modular homes for natural disaster zones.

瑞典国家铁路公司(SJ)是将预制胶合木材料应用在瑞典各地大厅和站台的框架结构中的早期实践者。在马尔默、斯德哥尔摩、桑兹瓦尔和哥德堡等地建造的优雅屋顶至今仍在使用。位于韦斯特拉斯的一座建造于1937年飞机机库的木质屋顶长期保持着世界上最大跨度的纪录,其跨度达到55米。

人们对具有创造性和审美吸引力的CLT和胶合木屋顶结构的兴趣正在全球范围内增长。

瑞典木材公司(Swedish Wood)的结构工程师Eric Borgstrom说:"由于其强度和重量轻的特点,木材非常适合承载大跨度的结构,而且通常很壮观。"

日本建筑师坂茂(Shigeru Ban)对有利于气候的结构框架和屋顶形式的木结构设计进行试验时表现突出。在2014年,他被授予著名的"普利兹克奖";这个奖通常被称为"建筑世界的诺贝尔奖"。他获奖作品有美国的阿斯彭艺术博物馆项目和法国的Pompidou-Metz中心项目的翻腾木制屋顶,更不用说他的应对自然灾害的临时模块式房屋。

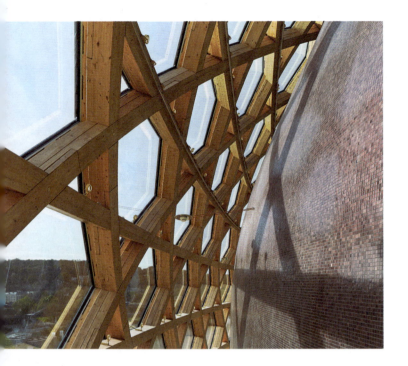

Case 1: La Seine Musicale

Architect: Shigeru Ban Architects/Jean de Gastines Architectes
Client: City of Boulogne-Billancourt
Structural engineer: Hess Timber
Cost: 170 miljoner euro
Size: 36500 m²

The island of Seguin sits in the River Seine, where it runs through the Paris suburb of Boulogne-Billancourt. It used to be home to a large car plant, but when Renault closed the factory, the buildings were demolished and plans for the island's future began to be hatched. Now, in addition to an office complex, the site has a magnificent egg-shaped building rising up from austere concrete foundations. La Seine Musicale was designed by Shigeru Ban Architects in collaboration with local architect Jean de Gastines.

Case 1: La Seine Musicale
案例1：法国塞纳河音乐厅

案例1：法国塞纳河音乐厅

建筑师：Shigeru Ban Architects/Jean de Gastines Architectes 建筑事务所
客户：布洛涅－比扬古市
结构工程师：Hess Timber 公司
花费：1.7 亿欧元
规模：36500 平方米

塞金岛位于塞纳河中部，流经巴黎郊区的布洛涅－比扬古。它曾经是一个大型汽车厂的所在地，但当雷诺关闭了工厂，这些建筑被拆除，并为岛上未来的计划做准备。现在，除了一个办公综合体，这块场地还有一个从简洁的混凝土基座上建起的宏伟的蛋形建筑。塞纳河音乐厅是由 Shigeru Ban 建筑事务所指派与当地建筑师 Jean de Gastines 合作设计的建筑项目。

La Seine Musicale is a multicultural meeting place with flexible and generously appointed spaces for shops, clubs and rehearsals, and for all sorts of events, including musicals, concerts and conferences. There is capacity for 4000 people seated or 6000 standing. The project's big number is actually a small concert hall, with 1150 seats, which is the home stage of the Insula Symphonic Orchestra. Here exposed wood of various shapes and origins is a commanding presence.

Externally, the building is like a huge bird's nest, built around an egg-shaped frame that forms a strong shell structure in spruce and beech glulam. The rounded form of the glulam was achieved using computer-controlled clamps. The entire structure is clearly visible through the glazed façade.

The advanced joinery technology has echoes of traditional Japanese wooden architecture. In addition, the whole structure can be taken apart and re-assembled without damaging the material. A "sail" of solar panels wraps around the outside of the bird's nest and moves on rails to follow the movement of the sun, while also shading the lobby from direct sunlight.

The interior has oak floors, the stage is made from cedar and the decorative ceiling comprises 916 wooden hexagons filled with 28000 cardboard tubes in four different sizes. To improve the acoustics, a glistening web of birch plywood and oak veneer strips lines the walls. The effect is stunning, exciting and surprisingly expressive.

塞纳河音乐厅是一个多元文化的会议场所，也为商店、俱乐部和排练以及各种活动（包括音乐剧、音乐会和会议）提供了灵活和设备齐全的室内场地。这个场地可容纳 4000 人的座椅，或 6000 人站立。这个项目的主体部分实际上是一个小音乐厅，拥有 1150 个座位，这是岛内交响乐团的主场。在这个项目中，各种形状的暴露式的原木结构是建筑的一大亮点。

从外部看，这座建筑就像一个巨大的鸟巢，围绕着一个蛋形框架建造，形成了一个由云杉和山毛榉胶合木搭建的坚固的壳体结构。它采用计算机控制夹具，建构了圆形的胶合木构件。整个结构形式在玻璃的建筑表皮下清晰可见。

先进的细木工技术与日本传统木结构建筑相呼应。此外，整个结构可以拆卸和重新组装而不损坏材料本身。由太阳能电池板组成的"帆"环绕在鸟巢的外面，沿着轨道以跟随太阳的移动而移动，同时也遮挡了直射大厅的阳光。

室内铺有橡木地板，舞台由雪松制成，装饰性天花板由 916 个木制六边形组成，其中填充了 28000 个 4 种不同尺寸的硬纸板管。为了改善音响效果，桦木胶合板和橡木条组成的闪闪发光格栅排列在墙壁上。效果是惊人的、令人兴奋和惊艳的。

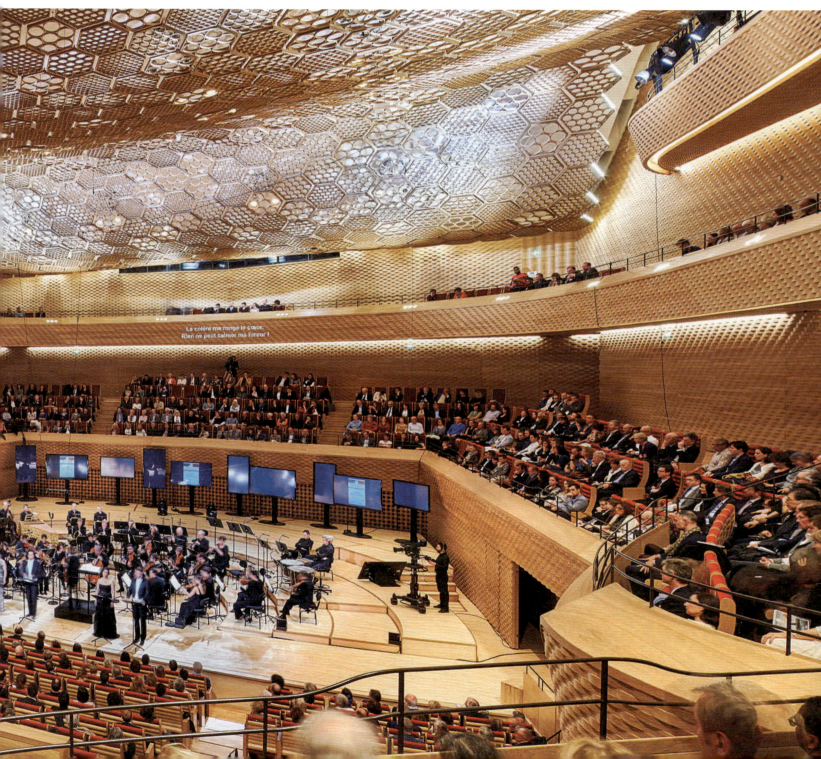

Case 2: Alice Milliat Sports Hall

Architect: Dietrich Untertrifaller Architecten/Tekhnê Architectes
Client: City of Lyon
Structural engineer: Société Dauphinoise Charpente Couverture, sdcc
Size: 2680 m²
Cost: 523 miljoner euro

Lyon has another interesting example of how prefabricated glulam beams can be used to bridge large spans. It's on a smaller scale than in the expressive La Seine Musicale, but it's no less innovative. In a social development project in Bon Lait, near the Rhône river on the outskirts of Lyon, a newly built sports hall forms the hub around which the project revolves. The building is designed not only as a sports hall for the school's pupils, but equally as a meeting place for the local community.

案例2：法国爱丽丝米丽叶运动馆

建筑师：Dietrich Untertrifaller Architecten/Tekhnê Architectes 建筑事务所

客户：里昂市

结构工程师：Société Dauphinoise Charpente Couverture, sdcc 公司

规模：2680 平方米

花费：5.23 亿欧元

在里昂有另一个关于预制胶合木梁如何搭建大跨度建筑的有趣例子。它的规模比富有表现力的塞纳河音乐厅的规模小，但它的创新性却丝毫不减。在里昂市郊罗纳河附近的邦莱特的一个社会发展项目中，一个新建的体育馆构成了项目的中心，整个项目围绕着它进行深化。该建筑不仅是学校学生的运动馆，同时也是当地社区的集会场所。

Case 2: Alice Milliat Sports Hall
案例 2：法国爱丽丝米丽叶运动馆

The sports hall, named after Alice Milliat, a French athlete and a champion of women's sport whose activism led to the introduction of women's events at the Olympic Games, is open to everyone and can be split into three separate halls for different activities.

According to the two architectural practices Dietrich Untertrifaller and Tekhnê Architectes, the building is designed as a simple wooden box with low thresholds and plenty of windows that open up onto the local square, stimulating activity that flows between the inside and the outside.

With a wooden frame and a façade of silvered larch, the building is as simple and accessible as the architects intended, despite its floor space of almost 3000 square metre and height of 9 metres. The fact that the insulation is made from straw packed into CLT boxes also adds an ecological dimension that goes beyond the ordinary. The flat roof is constructed from large glulam beams that are joined to the wall studs, which are also glulam. Between the roof beams are 18 sloping roof lights, shaped like CLT pyramids that face north and cast light over the 45 x 24 meter playing area.

With what the architects describe as a considered interaction between material choices, economics and design, in Bon Lait they have managed to achieve an aesthetically attractive building that also makes a clear statement about the role that wooden buildings can play in the modern French suburb. Their work also won them the 2017 Prix National de la Construction Bois, a national award for innovative wood construction projects in France.

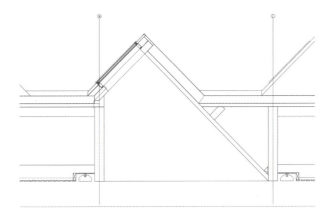

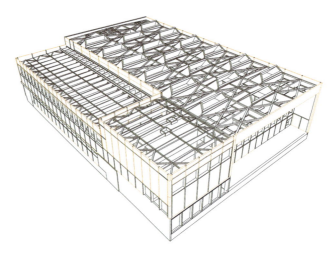

该体育馆以法国女子运动冠军爱丽丝米丽叶的名字命名，米丽叶积极行动，将女子项目带入奥运会场上。这个体育馆对每个公众开放，它可以分成3个不同的厅以举办不同的活动。

根据 Dietrich Untertrifaller 和 Tekhne Architectes 两家建筑事务所的设计，该建筑被设计为一个简单的木质盒子，有着低矮的入口和大量开向本地广场的窗户，以此来增加室内室外的交流活动。

该建筑由木质框架和银色落叶松立面装饰板组成，尽管该建筑面积近3000平方米，高度为9米，但正如建筑师所希望的那样，它是简洁和平易近人的。稻草被填充到 CLT 墙体中作为保温材料，这也增加了一种超越普通做法的生态维度。平屋顶是由与胶合木墙柱连接的大型胶合木梁构成的。屋顶横梁之间是18个倾斜的屋顶灯，形状像面朝北的 CLT 金字塔，向 45 米 ×24 米的游乐区投射光线。

建筑师将其描述为材料选择、经济性和设计之间经过思考的相互作用的产物，在邦莱特，他们成功地设计了一座具有美学吸引力的建筑，这也清楚地说明了木构建筑在现代法国郊区可以发挥作用。他们的作品还为他们赢得了2017年法国国家建筑博伊斯大奖赛（Prix National de la Construction Bois），这是法国创新木建筑项目的国家奖。

Case 3: Diamond Domes
案例 3：瑞士钻石穹顶网球场

Case 3: Diamond Domes

Architect: Rűssli Architekten
Client: Der ARGE Diamant, Medava + Partner
Structural engineer: Besmer-Brunner/ Neue Holzbau
Size: 1650.m²
Cost: 15.1 miljoner schweizerfranc

In many ways a more spectacular project, with a standout roof structure, is the Diamond Domes complex that forms part of Bürgenstock Hotels & Resort in Switzerland, 874 metres above sea level. The history of this spa hotel begins in 1864 with Franz Josef Bucher and carpenter Joseph Durrer opening a little sawmill that some years later expands into manufacturing wooden houses and wood flooring. Franz Joseph Bucher quickly realised that the area around Lake Lucerne was attractive to visitors and decided to build a hotel on Bürgenberg, a mountain since renamed Bürgenstock.

案例 3：瑞典钻石穹顶网球场

建筑师：Rüssli Architekten. 建筑事务所

客户：Der ARGE Diamant, Medava + Partner

结构工程师：Besmer-Brunner/ Neue Holzbau

规模：1650 平方米
花费：1510 万瑞士币

钻石穹顶网球场项目在许多方面，都是一个非常壮观的项目，它有着一个很出彩的屋顶结构。它作为瑞士 Bürgenstock 酒店和度假村的一部分，其海拔为 874 米。这家温泉酒店的历史始于 1864 年，当时 Franz Josef Bucher 和木匠 Joseph Durrer 开了一家小锯木厂，几年之后，这家锯木厂的业务扩展到了制造木屋和木地板。Franz Josef Bucher 很快意识到卢塞恩湖周围的地区对游客很有吸引力，于是他决定在伯根伯格山上建一家酒店，这座山后来改名为伯根斯托克山。

Right from the start, the hotel was an exclusive and popular haunt of the rich and famous. In addition to views of the Alpine peaks around Lake Lucerne, attractions included every imaginable treatment for increased well-being. Steam bath, sauna, massage, fitness – it was all there. Now there is also a new tennis and events hall built on a steep slope next to Bürgenstock. Or rather two halls with tent-like roofs, separated by an additional tennis court that in the winter converts into an ice skating rink. Both buildings, which are mirror images of each other, have an unusual roof structure that imitates the polygonal form of a rock crystal, hence the name Diamond Domes.

It was when the whole hotel complex was due for modernisation that Rüssli Architekten in Lucerne was commissioned to design the two tennis halls. The plan was to build the load-bearing structure in steel, but because of steel's movement in major temperature variations, the engineers at Neue Holzbau recommended using glulam for this particular project instead, with just a top plate of steel that also serves as a "ring beam". In just four months, all the construction material was prefabricated and transported by specialist vehicle to the construction site along narrow and nerve-racking alpine roads. By the end of another four months, five fitters had assembled the roof. It then took another two months to complete the interior.

"With all its different angles, varying heights above ground and the forces it has to deal with, the freestanding roof structure was a complex collaboration between engineers and assembly planners." states Neue Holzbau's project manager Thomas Steiner.

It required detailed solutions that met all the specifications and were practical when it came to assembling the structure's 700 glulam components and 300 sections of cross laminated timber that form the roof panels in each tennis hall. The CLT creates a stiff roof slab and stabilises the underlying roof structure, which comprises primary and secondary beams plus glulam noggings, forming a rhomboid grid that ingeniously disperses the load into the structure below.

从一开始,这家酒店是富人和名人的专属居所。除了欣赏卢塞恩湖周围的高山山峰外,景点还包括所有可以想象到的增加幸福感的治疗方法。蒸汽浴、桑拿、按摩和健身都在那里。现在,在伯根斯托克旁边的陡坡上,还修建了一座新的网球和比赛场馆。或者更确切地说,是两座有着像帐篷一样的屋顶的大厅,它们被另一个网球场隔开。这个网球场在冬天会变成溜冰场。这两座建筑互为镜像关系,有着不同寻常的屋顶结构。其结构模仿了岩石晶体的多边形形式,因此得名为钻石穹顶。

当整个酒店建筑群要进行现代化改造时,位于卢塞恩的 Russli Architekten 建筑事务所被委托设计这两个网球场。该计划原本是用钢建造承重结构的,但由于钢构件在温度变化的情况下产生形变,Neue Holzbau 的工程师建议在这个特殊的项目中使用胶合木,只在顶部使用一块钢板,同时也充当"环形梁"。在短短四个月的时间里,所有的建筑材料都是预制的,并由专业车辆沿着狭窄而令人紧张的高山公路运送到施工现场。又过了四个月,五个钳工把屋顶组装好了。然后又花了两个月的时间来完成内部装修。

Neue Holzbau 的项目经理 Thomas Steiner 说:"屋顶的独立式结构,其结构件有着不同的角度,离开地面有着不同的高度,承载着不同的应力。这个结构的设计是由工程师和装配规划师之间进行过复杂协作后完成的。"

它需要满足所有规范中对节点设计的要求,并且在装配 700 个胶合木结构组件和搭建每个网球场屋顶板的 300 块 CLT 板时也是实用的。CLT 创造了一个坚固的屋顶楼板,并稳固了屋顶下部的主梁和次梁以及胶合木钉结构,形成一个菱形网格,巧妙地将荷载分散到下面的结构中去。

Thomas Steiner stresses that masses of creativity and skill were vital throughout the process, which also placed major strength demands on the primary and secondary beams. A good example is the redesign of an existing patented fixing solution meant for just two beams, so that it instead handles five beams on different planes that meet at one node.

"The design of the roof meant we couldn't use traditional fixings. They needed to be developed as the work went on, preferably without using visible steel details, since the whole structure is exposed." says Thomas.

As well as the roof's different heights, they had to contend with the requirements for an exact fit with tiny tolerances when joining up the roof's rhomboid framework. They also had to take account of the movement in the wood, since on three sides the roof is surrounded by stiff concrete walls, while one side is a glass façade. The whole assembly was carried out on four scaffolding towers that were set up in each tower ahead of time.

The finished result is undeniably eye-catching and would certainly have pleased the father of glulam, master carpenter Otto Hetzer, over 100 years later.

In Sweden, the elliptical glulam beams still support the roof above Stockholm Central Station, built by the company Fribärande Träkonstruktioner (now Moelven Töreboda). But there has not really been anything on the domestic front that can measure up against Diamond Domes or La Seine Musicale. Eric Borgström believes there is interest in really big, monumental wooden roofs in Sweden, and he stresses that the wood industry is keen to take on such challenges, but it seems that foreign clients tend to be more interested in making a bold statement with unique structures.

Thomas Steiner 强调,在整个设计过程中,大量地投入创造力和技术是至关重要的,同时也对主梁和次梁提出了很高的强度要求。一个很好的例子是对现有的专利方案(仅针对两根梁)进行重新设计,从而使得不同平面方向上的五根梁可以相交在一个节点上。

"屋顶的设计意味着我们不能使用传统的连接件。他们需要随着工作的进行而升级。因为整个结构都是暴露的,所以最好不要使用可见的钢节点。"托马斯说。

除了屋顶构件的高差外,在连接屋顶菱形框架构件时,他们必须满足构件的微小公差的要求。他们还必须考虑到木材的收缩运动,因为屋顶的三面被坚硬的混凝土墙包围,而一面是玻璃幕墙。整个建筑是在四个安装塔上进行的,每个塔上都必须提前搭好脚手架。

完成后的作品无疑是引人注目的,肯定会让胶合木的父亲——木匠大师 Otto Hetzer 在 100 多年后的今天感到高兴。

在瑞典,椭圆胶合木梁仍然支撑着斯德哥尔摩中央车站上方的屋顶,该车站由 Fribarande Trakonstruktioner 公司建造(现在是 Moelven Toreboda)。但在国内,确实还没有任何建筑能与钻石穹顶网球场或塞纳河音乐厅相比。Eric Borgstrom 认为瑞典人对巨大的、具有纪念意义的木结构屋顶很感兴趣。他强调,木材行业渴望迎接这样的挑战,但似乎外国客户对其独特的结构形式更感兴趣。

世界大跨度木结构建筑一览表（2000—2020）
List of World Large Span Wood Structural Buildings from 2000 to 2020

2000	汉诺威世博会大屋顶 EXPODACH for Expo 2000 in Hannover	地点：德国 跨度：40米，高度20米 整个结构像是一个巨型的雕塑，它们与下方的水道和人工岛屿一起形成了一个建筑整体。	
2004	西温哥华水上运动中心 West Vancouver Aquatic Centre	地点：加拿大 该水上运动中心有26年的历史，包括了一个温水休闲游泳池，20米的大型滑水梯，家庭更衣室，多功能室，健身区以及公共观赏区。该水上运动中心因其独特的木制屋顶和木墙结构闻名，隐形的连接件允许结构线条连续流动，从而产生运动美观感，而这就恰恰强调了整体形式和建筑材料而非其连接件。屋顶结构曲梁和V形圆柱作为框架，它们之间的连接采用的是欧洲技术。每个连接由钢板焊接钢中嵌入销钉，即嵌入的两个地方采用环氧树脂粘在一起的。	
2010	鳟鱼湖溜冰场 Trout Lake Ice Rink	地点：加拿大 该冰场是用于参加2010年奥运会和冬季残奥会的参赛者练习用的，在赛事结束后该设施对公众开放。该建筑坐落于朝向东的斜坡的山坡脚下，在公园边与鳟鱼湖之间。这一位置与屋顶侧边轮廓一起，形成一个很浅的拱顶，犹如雨后春笋般从低围墙处立起——从而最大限度地减少了溜冰场内传出的超大音量的影响，同时，这样所形成的规模与周围的独栋邻里相匹配。	
2010	珀西诺曼水上运动中心 Percy Norman Aquatic Centre	地点：加拿大 珀西诺曼水上运动中心是温哥华希尔克雷斯特公园里新莱利公园社区中心的一部分。在2010年奥运会和冬季残奥会期间，该社区中心成为当时冰壶比赛场馆，同时游泳池成为运动员训练的区域。社区中心的主要钢筋结构通过中枢前庭连接到胶合木与重型木构造的游泳池，在中枢前庭处保留着现有的穿越公园的一条人行小道。这6200平方米的水上活动中心将包括一个休闲池、50米竞赛泳池和一个室外水上运动设施。	

2010	斯阔米什游客探险中心 Squamish Adventure Centre	地点：加拿大 建筑呈椭圆形，屋顶犹如巨型的翅膀盘旋于透明的墙体上方，该设计旨在为游客营造一片海岸山脉般浩渺的情怀。客户群体希望就地取材，以当地材料建造一栋引人注目的建筑，同时也为即将进行的海滨重建项目设置一个建筑质量基准。中心的顶部设计为弯曲的蝴蝶形屋顶，轻轻地盖在支撑结构的上方，支撑结构由一些裸露的木柱、支架和横梁构成。该蝴蝶形屋顶由 35 根不同的复合钢和木桁架组成，每一个木桁架都有各自独特的几何形状。	 
2010	温哥华会展中心 Vancouver Convention & Exhibition Centre	地点：加拿大 温哥华会展中心扩展项目面积为 100000 平方米，获得了 LEED 铂金认证（为该绿色建筑评估体系最高认证级别）。作为 2010 年冬季奥运会和冬季残奥会的传播中心，这里成为卑诗省林产工业的展示场所。整个建筑建立在 1000 根桩子和一个混凝土平台上，通过这种方式，建筑的钢结构就能够连接道路和铁轨。会展中心的展示空间面积为 90 米 ×225 米，同时带有一个面积 27 米 ×36 米的结构栅格。建筑的最大屋顶净跨度为 55 米。	
2010	梅茨蓬皮杜木结构文化中心 Centre Pompidou-Metz	地点：法国 跨度：80 米 屋顶表面积为 8000 平方米，由 16 千米的胶合层压木材组成，相交形成类似于中国竹编草帽的六边形木制单元。 屋顶的几何形状是不规则的，且在整个建筑物上布有曲线和反曲线，特别是三个展厅。整个木结构覆盖着白色玻璃纤维膜和聚四氟乙烯涂层，具有自清洁的特点，避免阳光直射，同时在夜间提供透明视图。	 

2010	赫里斯九桥高尔夫俱乐部 Haesley Nine Bridges Golf Clubhouse, Yeoju, South Korea	地点：韩国 跨度：40米 参数化设计，现代编织形式结构。六边形网状结构，从天花到柱子完全由六边形的木肋编织而成，柱子仿佛是网状结构的突变形成的，同时具有美感，让人联想到四周连绵的群山。	
2011	大都会太阳伞 Metropol Parasol	地点：西班牙塞维利亚 J. Mayer H. Architects 建筑事务所在西班牙塞维利亚设计的大都市太阳伞于2011年竣工。这是一个由混凝土和木材组成的巨大伞形结构，它将成为塞维利亚新的城市中心，作为一个地标建筑，它象征了城市的文化内涵。	
2012	赫尔辛基动物园观景塔 Kupla – Helsinki Zoo Lookout tower B	地点：芬兰 跨度：4~8米 高度：10米 展馆位于海拔18米的悬崖顶上，可以欣赏到壮观的海景和赫尔辛基市，10米高的塔楼是Korkeasari岛一个精致透明的地标。其自由形态的灵感来源于它的自然环境。承重的网壳结构由72个胶合木条组成，由7种预制形式的60毫米×6毫米木材段现场弯曲和扭曲而成。超过600个螺栓接头将壳体结构固定在一起。	
2015	苏州园博会主展馆 Main Hall in Horticultural Exposition of Jiangsu Province in Suzhou	地点：中国 跨度：45米 主展馆为钢木混合结构，主体外围支撑体系为钢结构，屋面为网壳结构，采用胶和木作为主次梁，网格式分布；主梁的宽度最长可达37米，高度在8米。	

2016	卡尔加里基督教青年会中心 Remington YMCA	地点：加拿大 YMCA（Young Men's Christian Association 基督教青年会）是加拿大西部最新、最吸引人的健康设施之一，服务于 Calgary 市及周边地区的居民，主导该地区的健康服务和社区精神建设。YMCA 宽阔的玻璃幕墙与明亮开放的空间产生了一种联系感，以引起居民的兴趣并鼓励他们参与。作为一个新城市社区的繁华中心，该设施体现了 Calgary 市建设充满活力社区的战略，内部特色设施有休闲场地、比赛专用泳池、热水浴池、桑拿房、室内体育场、跑步道、健身房、日托、儿童看护和公共图书馆。	
2016	克拉马特体育中心 Clamart Sport Centre	地点：法国 跨度：40 米 新的克拉马特体育中心是一座具有真实意义上的体育运动之城，其外形随着风景的曲线变化而变化。技术解决方案是将立面和屋顶连成一个连续的结构体系。该运动中心包括体育馆、武术道场、田径场和网球场，全部位于建筑的一层。	
2016	Ameublements Tanguay 零售店 Ameublements Tanguay CoArchitecture Levis, QC	地点：加拿大 Ameublements Tanguay 选择了木质结构作为其 55 年经营以来最大的零售店。该建筑位于 Trois-Rivières 的 55 区开发区，总表面积为 6960 平方米。它的设计特点是大跨度支撑托梁区域，第一部分的尺寸为 10 米 ×9.5 米，第二部分的尺寸为 12.5 米 ×15 米。前后两部分通过中庭相连，在展示木质结构的同时，提供室内充足的光线。	
2017	西弗雷泽曲棍球比赛中心 West Fraser Centre	地点：加拿大 林业是 Quesnel 经济的重要组成部分，因此社区领导人很自然地将木材作为其新竞技场的关键结构和饰面材料，用以促进该地区的木材文化。这座两层楼的建筑是社区的焦点，它已成为集现场娱乐、贸易展览和社区活动于一体的重要场所。	

| 2017 | 贵州省榕江室内游泳馆 Swimming pool in Rongjiang, Guizhou Province | 地点：中国 |

贵州省榕江室内游泳馆位于贵州省黔东南苗族侗族自治州榕江县，内设 50 米 × 50 米正式比赛池和 25 米 × 25 米训练池，总建筑面积 11455 平方米，占地面积 6180 平方米，建筑地下一层，地上两层，建筑高度为 20.05 米。建筑设计使用年限为 50 年，建筑结构安全等级为一级。建筑地下室及一层采用混凝土框架体系，二层以上采用木结构体系。

游泳馆中部花桥和鼓楼采用传统木结构，充分体现了民族特色和地域特点。泳池上部屋盖采用张弦木拱体系，跨度 50.4 米，为国内跨度第一和面积第一的现代木结构屋盖。木拱为 2 毫米 ×170 毫米 ×1000 毫米双拼胶合木构件，沿弧长三段拼接。木拱采用 6 根木撑杆与主索形成张弦结构，并与纵向索和屋面索形成完整稳定体系。自平衡的张弦木拱支承于滑移支座，消除支座水平推力，有效地降低了造价。采用木结构与玻璃形成的"天河"结构悬挂于主拱。

| 2017 | 德尔布鲁克社区康乐中心 Delbrook Community Recreation Centre | 地点：加拿大 |

该建筑有朝北的小型空间和朝南的大型空间组织。互联的循环"脊柱"将建筑物一分为主要组织元素。主要空间包括：室内，25 米长的游泳池和休闲游泳池，成人和康复计划（带轮椅升降机），运动班工作室（带听觉障碍的感应环路地图），健身中心和健身房。其他空间包括球拍球场，艺术工作室，陶艺工作室，青年、高级和幼儿中心以及会议室。大堂强调户外连接，这是山腰社区的特色，设有咖啡厅和座位。

| 2017 | 基督教青年会的谢恩之家康乐中心 Shane Homes YMCA At Rocky Ridge | 地点：加拿大 |

谢恩之家康乐中心是世界上最大的基督教青年会中心之一。它为该地区提供基本的健乐设施，是社区聚会、文化节目、娱乐和儿童保育服务的区域中心。

2018	Tsleil Wautututh 管理和医疗康复中心 Tsleil-Waututh Administration & Health Care	地点：加拿大 该建筑的主要结构由钉合的 SPF 板组成，并支撑在花旗松胶合木（GLT）梁上。天花板的曲线是由各种不同长度的层压花旗松板构成。墙壁是用常规的轻木结构建造的。同时，大雪松木板用于主要集会区，将传统与当代设计融为一体。	
2018	明纳健康活动中心 Minoru Centre for Active Living	地点：加拿大 位于 Minoru 的老年人活动中心是 55 岁及以上老年人都可使用的设施，它为老年人提供了体育运动、健康和社交的生活方式。 明纳健康活动中心，包括一个宽敞明亮的 8500 平方英尺健身中心。各种机器和设备均可满足各个年龄段和身体素质的用户。	
2019	BUGA 木展厅 BUGA Wood Pavilion	地点：德国 BUGA 木展馆采用数字化木材建造的新方法。它的分段木壳结构借鉴了海胆板状骨架中发现的生物学原理。作为该项目的一部分，设计师开发了一个机器人制造平台，用于自动组装和铣削展厅的 376 个定制空心木构件。这种制造过程可确保所有构件以亚毫米级的精度装配在一起，就像一个大的三维拼图。令人惊叹的木制屋顶跨度超过 BUGA 的主要活动和音乐会场地之一，长达 30 米，使用了最少的材料，同时还产生了独特的建筑空间。	 

注：此表罗列案例不包含《木建筑》第 1 辑和第 2 辑中已介绍的大跨木结构建筑案例。

《木建筑》使用内容授权书

Letter of Authorization
授权书

This letter is to certify that Timber Architecture Research Center of School of Design, Shanghai Jiao Tong University, is entitled to use the selected original articles and pictures for the book Mù Architecture.

本授权书声明，上海交通大学设计学院木建筑研究中心有权在《Mù Architecture（木建筑）》一书中使用被选的文字和图片。

Canada Wood has supported the preparation and publication of Mù Architecture by providing articles and pictures of modern timber buildings since 2018.

加拿大木业自 2018 年以来，一直为《Mù Architecture（木建筑）》一书提供现代木结构建筑的相关文章和图片支持。

All the original articles and pictures can only be used in China by Timber Architecture Research Center of School of Design, Shanghai Jiao Tong University, for Mù Architecture. China Architecture & Building Press is the designated press who is entitled to use the original articles and pictures only for the publication of Mù Architecture.

在中国，所有被选的文章和图片，仅上海交通大学设计学院木建筑研究中心有权用于《Mù Architecture（木建筑）》一书。中国建筑工业出版社作为该书的指定出版社，可以使用被选的文章和图片，用于《Mù Architecture（木建筑）》一书的出版。

All the selected articles and pictures used in Mù Architecture are listed in the appendix.

所有被选的文章和图片详见附件列表。

Date （日期）: 10.30.2019

Signature（签字）:

Appendix: Selected articles and pictures used in Mù Architecture
附录：所有在《Mù Architecture （木建筑）》一书中使用的被选文章和图片。

Projects	Source from
Surrey Central City	https://www.naturallywood.com/wood-design/project-gallery/surrey-central-city
Richmond Olympic Oval	https://www.naturallywood.com/resources/richmond-olympic-oval Photographers: Craig Carmichael, KK Law, Stephanie Tracey, Martin Tessler
Grandview Heights Aquatic Centre	https://www.naturallywood.com/wood-design/project-gallery/grandview-heights-aquatic-centre Images courtesy of HCMA Architecture + Design
西郊宾馆意境园多功能厅	加拿大木业 (Canada Wood) 上海绿建建筑装饰设计有限公司 (Green-A Architecture & Decoration Design Co., Ltd.) 苏州昆仑绿建木结构科技股份有限公司 (Suzhou Crownhomes Co., Ltd.)
Kwakiutl Wagalus School	https://www.naturallywood.com/wood-design/project-gallery/kwakiutl-wagalus-school Photo courtesy of Lubor Trubka Associates | Peter Powles Photography
Pacific Autism family Centre	https://www.naturallywood.com/wood-design/project-gallery/pacific-autism-family-centre Photo credit: Derek Lepper
龙湖两江长滩原麓社区中心	上海成执建筑设计有限公司 (Challenge Design)
上海西岸人工智能峰会 B 馆建造实践	上海创盟国际建筑设计有限公司 (Archi-Union Architects)
StructureCraft Manufacturing Facility	https://www.naturallywood.com/wood-design/project-gallery/structurecraft-manufacturing-facility Photo courtesy of StructureCraft Builders
Wood Innovation Research Lab	https://www.naturallywood.com/resources/wood-innovation-research-lab Photo courtesy of Michael Elkan Photography Photo courtesy of University of Northern British Columbia

Letter of Authorization
授权书

This letter is to certify that Timber Architecture Research Center of School of Design, Shanghai Jiao Tong University, is entitled to use the selected original articles and pictures for the book Mù Architecture.

本授权书声明，上海交通大学设计学院木建筑研究中心有权在《Mù Architecture （木建筑）》一书中使用被选的文字和图片。

Daxing'anling Arctic pole woodIndustry.co.Ltd Bai Weidong has supported the preparation and publication of Mù Architecture by providing articles and pictures of modern timber buildings since 2019.

大兴安岭神州北极木业有限公司　白伟东 自 2019 年以来，一直为《Mù Architecture （木建筑）》一书提供现代木结构建筑的相关文章和图片支持。

All the original articles and pictures can only be used in China by Timber Architecture Research Center of School of Design, Shanghai Jiao Tong University, for Mù Architecture. China Architecture & Building Press is the designated press who is entitled to use the original articles and pictures only for the publication of Mù Architecture.

在中国，所有被选的文章和图片，仅上海交通大学设计学院木建筑研究中心有权用于《Mù Architecture （木建筑）》一书。中国建筑工业出版社作为该书的指定出版社，可以使用被选的文章和图片，用于

《Mù Architecture （木建筑）》一书的出版。

All the selected articles and pictures used in Mù Architecture are listed in the appendix.

所有被选的文章和图片详见附件列表。

Date （日期）： 2020年3月1日

Signature （签字）： 白伟东

Appendix: Selected articles and pictures used in Mù Architecture

附录：所有在《Mù Architecture （木建筑）》一书中使用的被选文章和图片。

Projects	Source from
长春市全民健身活动中心游泳馆	Daxing'anling Arctic pole woodIndustry.co.Ltd Bai Weidong

Letter of Authorization

授权书

This letter is to certify that Timber Architecture Research Center of School of Design, Shanghai Jiao Tong University, is entitled to use the selected original articles and pictures for the book Mù Architecture.

本授权书声明，上海交通大学设计学院木建筑研究中心有权在《Mù Architecture（木建筑）》一书中使用被选的文字和图片。

 Shanghai Zhenyuan Timber Structures Engineering Co.,Ltd has supported the preparation and publication of Mù Architecture by providing articles and pictures of modern timber buildings since 2019.

   上海臻源木结构设计工程有限公司   自 2019 年以来，一直为《Mù Architecture（木建筑）》一书提供现代木结构建筑的相关文章和图片支持。

All the original articles and pictures can only be used in China by Timber Architecture Research Center of School of Design, Shanghai Jiao Tong University, for Mù Architecture. China Architecture & Building Press is the designated press who is entitled to use the original articles and pictures only for the publication of Mù Architecture.

在中国，所有被选的文章和图片，仅上海交通大学设计学院木建筑研究中心有权用于《Mù Architecture（木建筑）》一书。中国建筑工业出版社作为该书的指定出版社，可以使用被选的文章和图片，用于

《Mù Architecture （木建筑）》一书的出版。

All the selected articles and pictures used in Mù Architecture are listed in the appendix.

所有被选的文章和图片详见附件列表。

Date （日期）： 2020.3.4

Signature （签字）：

Appendix: Selected articles and pictures used in Mù Architecture

附录：所有在《Mù Architecture （木建筑）》一书中使用的被选文章和图片。

Projects	Source from
天津欢乐谷演艺中心	上海臻源木结构设计工程有限公司
贵州紫云自治县格凸河攀岩基地-观赛广场	上海臻源木结构设计工程有限公司
Vertical Screen Warminster Campus 美国竖屏沃敏斯特总部办公楼	上海臻源木结构设计工程有限公司

Letter of Authorization
授权书

This letter is to certify that Timber Architecture Research Center of School of Design, Shanghai Jiao Tong University, is entitled to use the selected original articles and pictures for the book Mù Architecture.

本授权书声明，上海交通大学设计学院木建筑研究中心有权在《Mù Architecture （木建筑)》一书中使用被选的文字和图片。

___Zhan hui___ has supported the preparation and publication of Mù Architecture by providing articles and pictures of modern timber buildings since 2019.

__詹晖__ 自 2019 年以来，一直为 《Mù Architecture （木建筑)》一书提供现代木结构建筑的相关文章和图片支持。

All the original articles and pictures can only be used in China by Timber Architecture Research Center of School of Design, Shanghai Jiao Tong University, for Mù Architecture. China Architecture & Building Press is the designated press who is entitled to use the original articles and pictures only for the publication of Mù Architecture.

在中国，所有被选的文章和图片，仅上海交通大学设计学院木建筑研究中心有权用于《Mù Architecture （木建筑)》一书。中国建筑工业出版社作为该书的指定出版社，可以使用被选的文章和图片，用于《Mù Architecture （木建筑)》一书的出版。

All the selected articles and pictures used in Mù Architecture are listed in the appendix.

所有被选的文章和图片详见附件列表。

Date （日期）: _____2020/3/3_____

Signature （签字）: _____

Appendix: Selected articles and pictures used in Mù Architecture

附录：所有在《Mù Architecture （木建筑）》一书中使用的被选文章和图片。

Projects	Source from
云南弥勒太平湖森林小镇国际木屋会议中心	詹晖

Letter of Authorization
授权书

This letter is to certify that Timber Architecture Research Center of School of Design, Shanghai Jiao Tong University, is entitled to use the selected original articles and pictures for the book Mù Architecture.

本授权书声明，上海交通大学设计学院木建筑研究中心有权在《Mù Architecture（木建筑）》一书中使用被选的文字和图片。

<u>Yiju Ding</u> has supported the preparation and publication of Mù Architecture by providing articles and pictures of modern timber buildings since 2019.

<u>丁一巨</u> 自2019年以来，一直为《Mù Architecture（木建筑）》一书提供现代木结构建筑的相关文章和图片支持。

All the original articles and pictures can only be used in China by Timber Architecture Research Center of School of Design, Shanghai Jiao Tong University, for Mù Architecture. China Architecture & Building Press is the designated press who is entitled to use the original articles and pictures only for the publication of Mù Architecture.

在中国，所有被选的文章和图片，仅上海交通大学设计学院木建筑研究中心有权用于《Mù Architecture（木建筑）》一书。中国建筑工业出版社作为该书的指定出版社，可以使用被选的文章和图片，用于《Mù Architecture（木建筑）》一书的出版。

All the selected articles and pictures used in Mù Architecture are listed in the appendix.

所有被选的文章和图片详见附件列表。

Date （日期）: 2020.03.26

Signature （签字）: J-E

Appendix: Selected articles and pictures used in Mù Architecture

附录：所有在《Mù Architecture （木建筑）》一书中使用的被选文章和图片。

Projects	Source from
太原植物园	丁一巨

Letter of Authorization
授权书

This letter is to certify that Timber Architecture Research Center of School of Design, Shanghai Jiao Tong University, is entitled to use the selected original articles and pictures from Swedish Wood magazine Trä for the book Mù Architecture.

本授权书声明,上海交通大学设计学院木建筑研究中心有权在《Mù Architecture （木建筑)》一书中使用瑞典木业协会杂志《Trä》中的部分原始文字和图片。

Swedish Wood, member of European Wood, has been supporting the preparation and publication of Mù Architecture by providing articles and pictures of modern timber buildings from Swedish Wood magazine Trä since 2014.

瑞典木业协会作为欧洲木业协会的成员,自2014年以来,一直为《Mù Architecture （木建筑)》一书提供现代木结构建筑的相关文章和图片支持,文章和图片来源于瑞典木业协会杂志《Trä》。

All the original articles and pictures from Swedish Wood magazine Trä can only be used in China by Timber Architecture Research Center of School of Design, Shanghai Jiao Tong University, for Mù Architecture.

Appendix: Articles and pictures from Swedish Wood magazine Trä used in Mù Architecture

附录：所有在《Mù Architecture （木建筑)》一书使用的来源于瑞典木业协会杂志《Trä》的文章和图片。

	Title	Issue
1	Garage for the jetset	ISSUE 3, 2015
2	Football club gets a stadium in a league of its own	ISSUE 1, 2017
3	Flower garden with stepped roof	ISSUE 3, 2017
4	Simple from bears fruit	ISSUE 3, 2017
5	Vault welcomes tourists	ISSUE 4, 2018
6	pool with woodland views	ISSUE 1, 2018
7	Suspension roof structures create elegant spans	ISSUE 4, 2017
8	Natural material at the world's greenest airport	ISSUE 3, 2017
9	Ålgård new cultural meeting place	ISSUE 3, 2015
10	Statement pieces	ISSUE 2, 2018
11	Advanced curves based on mythology	ISSUE 1, 2019
12	Function and aesthetics thanks to meticulous planning	ISSUE 1, 2019

China Architecture & Building Press is the designated press who is entitled to use the original articles and pictures from Swedish Wood magazine Trä only for the publication of Mù Architecture.

在中国，所有来源于瑞典木业协会杂志《Trä》的文章和图片，仅上海交通大学设计学院木建筑研究中心有权用于《Mù Architecture（木建筑）》一书。中国建筑工业出版社作为该书的指定出版社，可以使用来源于瑞典木业协会杂志《Trä》的文章和图片，用于《Mù Architecture（木建筑）》一书的出版。

All the selected articles and pictures from Swedish Wood magazine Trä used in Mù Architecture are listed in the appendix.

所有来源于瑞士木业协会杂志《Trä》的文章和图片详见附件列表。

Date（日期）: 2020-04-27

Signature（签字）:

上海臻源木结构设计工程有限公司

上海臻源是一家专注于现代木结构建筑设计、制造及安装的专业公司，尤其擅长大跨度木结构公共建筑，上海臻源在上海及重庆两个区域设有办公室，在江苏南通及苏州设有加工制造基地，为客户提供个性木结构建筑定制的整体解决方案。

公司关注最新木结构发展动态，以独有的国际视野和严谨的态度，注重将传统技艺的经验累积和全球工业化数字化的新兴高效技术相融合。公司项目广泛应用于文旅建筑、体育馆、售楼中心、展览馆、文物建筑改造修缮、别墅建筑、树屋以及景观桥梁等。

公司拥有稳定的设计团队，采用专业的三维木结构模型设计方法，依托先进的工程结构分析技术，轻松驾驭复杂、突破想象力的木结构项目，给木结构建筑带来全新建造理念，拓展了木结构的应用范围。能够按照中国及北美相关规范标准进行木结构设计，定期与国际一流专业公司进行合作与交流，整合国际相关资源。中国公司自成立以来，持续为北美客户提供木结构设计服务。

公司拥有基于工业4.0理念的CAM全自动数控机床加工中心，以及智能化的多维加工机器人设备。设计与加工中心无缝对接，设计成果直接输入全自动数控机床，构件加工更精确、安全，减少现场安装误差。加工工厂获得国际第三方质量认证，制定了臻源独有的质量企业标准，拥有完善的质量控制体系。

公司拥有一支从业十多年的能够打硬仗的木结构项目管理及安装团队。在国内完成过大量高难度的大跨度木结构建筑的安装。

地址：上海市静安区共和新路2993号和源中环企业广场406室
网址：www.zytimber.com
电话：021-66057776（上海）
　　　023-63536936（重庆）
邮箱：1294299005@qq.com（上海）
　　　17620296@qq.com（重庆）
微信公众号：臻源木结构

河北怀来上谷水郡会所

云南西双版纳悦景庄

贵州紫云格凸河户外运动体验馆

上海融嘉木结构房屋工程有限公司

上海融嘉木结构房屋工程有限公司从2007年开始致力于木结构建筑，传承传统木结构工艺结合现代木结构建筑体系，历经多年的市场风雨洗礼和坚持不懈，目前，公司木结构建筑的各项技术和服务均达到行业领先水平。融嘉木屋具有自主设计、工厂预制、现场搭建施工一体化能力，拥有成套、专业的现代化加工设备，擅长建造大型梁柱场馆、接待中心、中高端私人别墅及木结构桥梁，并符合国家《十三五"装配式建筑行动方案"》，现有管理人员12位，设计人员8位，拥有丰富的经验和技术支持，每年可承建并安装60000m²的工程项目。融嘉秉承着"专注、专业、专心"的企业信仰，立足国内，扬帆国际市场，近年来创作了一系列有影响力的作品……

融嘉全资控股的"江苏惠优林集成建筑科技有限公司"是一家专注于集成装配式建筑的科技型企业，公司拥有13800m²的标准厂房，引进台湾和意大利先进的生产设备，可年产15000m³以上的大截面大规格胶合木，已经投产的胶合木生产线2条，CLT生产线1条，曲形梁生产线1条，并在业内率先采用高频设备。

江苏惠优林集成建筑科技有限公司拥有十余年的胶合木生产经验，并且拥有强大的设计深化、生产管理、施工安装团队，为客户提供全方位的木结构系统集成服务。公司重视产品的研发创新，与南京林业大学建立长期紧密的战略合作关系，是南京林业大学木结构系的产学研基地及研究生工作站，南京林业大学在公司生产基地设立木结构研究实验室及胶合木实验室，公司承接了南京林业大学木结构系多项木结构研究课题，并获得多项专利。

地址：上海市松江区九新公路2888号申新广场
　　　5号楼A-F座
网址：www.rongjiamuwu.com
电话：18621751581
邮箱：530393646@qq.com
微信公众号：融嘉绿建

四川泸州老窖品牌馆

上海亦境建筑景观有限公司

上海亦境建筑景观有限公司成立于 2008 年 8 月 8 日。以"和为贵、诚为实、新为特"为企业服务理念，致力于规划咨询、景观设计、建筑设计与景观工程施工。公司目前具备建筑行业和风景园林专项双甲级资质，通过了 ISO9001 质量管理体系认证、GB/T 28001 职业健康安全管理体系认证和 ISO14001 环境管理体系认证。

公司与上海交通大学设计学院紧密合作，联合国内外知名高校专家团队，打造"产学研用"一体的创新设计、工程智造和人才孵化的特色平台（首批全国示范性风景园林专业学位研究生联合培养基地）。构筑起从项目前期策划、整体规划、旅游规划、建筑设计、景观设计、室内设计到景观工程施工的成熟的、全过程、一体化服务团队，形成多学科交汇融合、多专业分工合作的操作模式。

多年来，亦境团队以技术整合和创新思维为引领，形成城市精细化整治设计、园区规划与建筑设计、文旅与休闲空间设计、中式园林与建筑设计、可持续景观设计与营造等特色板块。业务遍布长三角，远及东南亚。打造出一批有影响力、示范性和引领性的作品，获得规划设计类奖 40 余项，其中 IFLA 亚太卓越奖 4 项，英国 BALI 国际项目奖 2 项。塑造出"亦真亦善"的亦境作品、亦境服务和亦境品牌。

地址：上海市普陀区中江路 388 弄
　　　国盛中心 1 号楼 3001 室
网址：www.edging.sh.cn
电话：021-61677866
邮箱：la@edging.sh.cn
微信公众号：上海亦境

上海静安府

上海松江醉白池公园

江苏镇江古运河地段

扬州园林设计院有限公司

扬州园林设计院有限公司成立于 2008 年，前身为扬州古典园林建设公司规划设计部，是国家级非物质文化遗产"扬州园林营造技艺"传承机构之一，扬州园林的杰出代表。

公司依托蜀冈瘦西湖风景名胜区，深入研究扬州古典园林博大精深的文化内涵、独到的造园理论、师法自然的设计理念以及高超的造园技术，主持设计复原了一系列扬州著名景点以及境外项目，获得了国家级、省级多个奖项及社会各界的广泛认可。近年来，公司致力于古典园林事业发展的同时，践行市委市政府"扬州园林"品牌"走出去"的要求，积极拓外部市场，秉持"心系生态、致敬自然"的理念，通过一系列城市更新、生态修复、园林景观等项目设计落地，获得了多项省、市级以及专业协会奖项认可，积累了丰富的经验，使得"扬州园林"品牌美誉度进一步提升。

公司下设景观设计、规划建筑设计、古典营造研究室、图文新媒体中心 4 个设计部门；建筑师、规划师、景观工程师、平面设计师等专业技术人员 50 人，其中高级工程师 3 名，中级工程师 16 名，技术实力雄厚；拥有风景园林工程设计专项资质（乙级），文物保护工程设计资质（丙级）。

公司业务范围涵盖风景名胜区、旅游度假区、城市景观系统、美丽乡村、特色小镇等规划设计；城市道路、广场、公园、水系、人居环境等景观设计；城市更新、古城保护、城市双修、环境综合整治等城市生态环境提升设计；古典园林、仿古建筑、各类民用建筑设计；文物保护工程设计；城市家居、标识标牌、数字多媒体、平面设计。

江苏扬州瘦西湖高速入口

江苏扬州文昌路

江苏扬州国医馆

地址：扬州市邗江区鸿福路 8 号
网址：www.yzylsj.com
电话：0514-82930338
邮箱：yzyzlsjy@163.com
微信公众号：扬州园林设计院

东莞市英仁建材科技有限公司

英仁科技致力于为国内的木结构建筑、古木建筑、木制品提供全面的防火、防虫、防腐、防霉保护。

英仁革命性的木材保护技术来自于北美。英仁是一种环保的水性涂料，经过近 20 年在国际市场的推广、使用、研究和发展，符合北美、欧洲及中国环保涂料的标准要求。

英仁的木材保护涂料环保、无毒害和无腐蚀性，挥发性有机物几乎为零。该技术适用于任何种类的木材，不管是硬木还是软木。卓越的耐久性能，无论在室内还是户外，对木材、木制品的全效保护始终如一。技术在不改变木材的结构性能的情况下，能最大限度地保留木材的自然纹理和原色。

目前技术应用于多个国内具有较大影响力的木结构项目：海口市民游客中心、云南五朵金花剧院、长春电影博物馆、江苏园艺博览会主展馆、北京法国学校、长春全民健身游泳馆、常州淹城初中体育馆、重庆长嘉汇、上海虹桥国际展汇、上海西郊游客中心、太原植物园、天府国际会议中心等多个大型公建木结构项目。多个项目采用形式检测或现场见证检测方式，对使用英仁产品后的木制品进行防火性能测试，均满足《木结构设计标准》（GB 50005-2017）对木建筑材料的燃烧性能的强制要求，达到国家标准《建筑材料及制品燃烧性能分级》（GB 8624-2012）的 B1 级。

地址：广东省东莞市莞城街道万园东路金汇广场 A 座 3305 室
网址：www.instarteco.com
电话：0769-22615120
邮箱：info@instarteco.com
微信公众号：英仁科技

四川成都天府国际会议中心

四川成都天府国际会议中心

海南海口市民游客中心

云南大理五朵金花剧院

赫英木结构制造（天津）有限公司

赫英木结构制造（天津）有限公司是瑞士赫英建筑木结构工程有限公司与香港远东工程有限公司合作在国内投资的外资独资企业，主要从事建筑胶合木构件的设计生产、制作以至安装全过程的建筑木结构工程公司。

瑞士赫英建筑木结构工程公司是有着140多年历史、国际知名的建筑木结构生产制造商，具有超前的设计理念、创新能力、先进的技术和加工工艺，其生产的木结构建筑功能强大、寿命持久、外形美观，具有强大的市场竞争优势。公司使用可再生木材生产粘接木构件，在环境保护和节约能源方面作出了巨大贡献，满足了21世纪建筑发展趋势的需求。长期积累的经验和企业信誉度，使客户对公司的产品和服务充满信心。

赫英木结构制造（天津）有限公司引进德国木构件生产设备，是目前国内可生产最大型木构件的全自动生产线，购置有世界最宽尺寸木构件四面刨光机。公司拥有设计人员、生产人员及安装队伍。提供由设计、生产、安装直至交锁匙的建筑木结构工程一条龙服务。

浙江杭州香积寺藏经楼

浙江杭州香积寺大雄宝殿

浙江杭州香积寺寺前广场

地址：北京市朝阳区太阳宫中路12号
电话：010-84299118/84299008
传真：010-65810086
邮箱：haring2008@sina.com

大兴安岭神州北极木业有限公司

大兴安岭神州北极木业有限公司成立于 2010 年 12 月 28 日，坐落于漠河市工业园区，是由中央直属企业大兴安岭林业集团公司出资设立的国有独资企业。公司注册资金 1.74 亿元，是国内大型木结构建筑生产加工企业之一，拥有木结构工程施工企业一级资质。

公司采用优质落叶松原料，绿色环保材料设计、生产、建造的重型、大型钢木、梁柱、轻型结构体系的木结构建筑，绿色低碳、健康安全，被称为"会呼吸的房子"，深受客户喜爱。

公司拥有核心技术及产品、自主知识产权、24 项国家专利，获得了国家科技进步二等奖，首批国家林业重点龙头企业、首届生态文明·绿色发展领军企业、中国木结构标准化优秀企业、中国林产工业 30 周年突出贡献奖、中国林业产业诚信 AAAAA 等荣誉 30 余项。取得了欧盟 CE 认证、绿色工厂认证、高新技术企业认证、省级企业技术中心认证和质量、安全、环保体系认证。木结构建筑品质跻身国际先进行列，在国内拥有"十个唯一"，是国内木结构建筑行业的领军企业，被权威人士誉为中国高端木结构建筑的领导者。神州北极木屋是"中国家居综合实力 100 强品牌"，是国内唯一一家进入 100 强的木结构建筑企业。

公司秉承"以人为本、以技为魂、客户至尊、追求卓越"的经营及服务理念，真诚与国内外朋友洽谈合作，发展共赢！

地址：黑龙江省大兴安岭地区漠河县西林吉镇加漠公路 497 公里 +600 米（XD264）
网址：www.szbjdxal.com
电话：0457-2886477
　　　400-657-9116
邮箱：smgszb01@126.com

滨江湿地度假会所

大兴安岭度假会所

云南弥勒太平湖国际会议中心

湖南麓上住宅工业科技有限公司

湖南麓上住宅工业科技有限公司成立于2017年9月，位于益阳市桃江县经济开发区，是一家集现代木结构建筑研发、设计、生产、安装、维保于一体的装配式木结构企业。现有标准化厂房面积12000m²，展示体验园占地愈8000m²。公司拥有管理团队50余人，其中现代木结构专业技术人才20余人。公司下设营销中心、木结构研发中心、生产中心、仓储物流中心及培训中心。公司拥有木屋生产线两条，每年生产木结构100000m²以上。

公司以装配式木结构为核心，大力拓展现代木结构在旅游度假项目、汽车营地项目、旅游扶贫项目、美丽乡村项目、特色小镇项目以及乡村自建房领域的应用。先后承担了长沙浔龙河生态艺术小镇、永州周家大院景区、怀化靖州飞山景区、湘西乌龙山民俗体验区、邵阳武冈云山露营公园、怀化大峡谷景区、北京2019世界园艺博览会湖南馆、河南汤阴湿地公园等大型装配式木结构工程的设计、生产和施工。

地址：湖南省益阳市桃江县经开区金牛路标准化厂房1号、3号（总部）
湖南省长沙市开福区潮宗御苑C栋1503室
网址：www.lushanxiaoyu.com
电话：0731-85351000
　　　400-160-0008
微信公众号：麓上木屋

湖南怀化鹤城怀化心田度假别墅

湖南浏阳大围山国家森林公园唯山居休闲度假会所

湖南邵阳武冈云山露营公园游客接待中心

湖南永州周家大院游客接待中心

福州小米木屋建设工程有限公司

福州小米木屋建设工程有限公司是福建省木结构建筑产业化企业，也是省内少数一家专业专注于木屋木结构建筑设计、木结构建筑部件生产、木结构建筑施工安装的木结构装配式企业。目前公司在福州近郊拥有一座占地面积约 $5000m^2$ 的胶合木生产基地。公司现有管理人员约 30 人。是一个生产有规模、施工懂技术、售后有保障、合作重信用的企业经营团队。

公司是全国木结构行业联盟协会（CWSA）副会长单位，常年与各大高校保持校企合作的良性互动。是福建省内为数不多的一家同时掌握轻型木结构 2×4 工法技术、正交胶合木墙体结构建筑、大跨度胶合木梁柱结构建筑、胶合木曲形梁柱子结构的企业。

2018 年"618"期间，公司推出的"一栋说走就走的房子"——小米移动房车备受市场青睐，2019 年"618"创新成果交易展期间，小米木屋用 6 根大型变截面曲形胶合木搭建而成的 $300m^2$ 大型构筑展示空间，惊艳全馆！期间，福建电视台新闻频道、福州晚报头版、海峡都市报、东南早报、中国网、新浪网等十几家国内和省内主流媒体争相报道。省领导登上木屋房车参观后，给予了高度评价。

"一直被模仿，从未被超越"，作为木结构装配式的先行先试企业的小米木屋，在"木屋会呼吸、沐浴大自然"的产品理念的指引下，正以其孜孜以求的精神面貌，成为建筑新领域的翘楚。

地址：福州鼓楼区乌山西路 318 号鼓楼科技大厦 4 楼
网址：117640.sites.fuhai360.com:8088
电话：159-5910-0609
邮箱：1324568@qq.com
微信公众号：小米木屋

福建农林大学附属幼儿园胶合木结构教学楼